Oils, Detergents and Maintenance Specialties

OILS, DETERGENTS
and
MAINTENANCE SPECIALTIES

Volume 1

MATERIALS and PROCESSES

by

BENJAMIN LEVITT, F.A.I.C.

Consultant to
Soap and Related Industries

CHEMICAL PUBLISHING COMPANY, INC.
NEW YORK 1967

Oils, Detergents and Maintenance Specialties, Volume 1, Materials and Processes

ISBN: 978-0-8206-0232-5

Chemical Publishing Company:
www.chemical-publishing.com
www.chemicalpublishing.net

First Edition:

© **Chemical Publishing Company, Inc.** - New York 1967

Second Impression:

Chemical Publishing Company, Inc. - 2011

Printed in the United States of America

Preface

Presented in this book are quantities of data, gathered during many years in my practice of chemistry, either as resident chemist in manufacturing plants, or as a consultant to the soap and related industries.

As in my previous volume, *Oil, Fat, and Soap*, published in 1951, the reader will find the subject matter presented in a clear and concise manner, so that he may readily satisfy his interest without extensive reading.

The style is didactic and such that it may readily be understood by non-technical personnel. Involved theoretical and technical discussions have been avoided. As far as possible, references have been included at the end of each chapter.

Although much of the information contained in this work is well within my own experience, much of it has been gleaned also from manufacturers' technical bulletins or through private correspondence.

Many patents have been cited, merely as examples of products discussed; this is not to be construed as recommending the infringement of any such patents. The various formulations are presented in good faith, but no warranty is given, nor is freedom from any patent to be inferred. With a little ingenuity on the part of the reader, many useful and efficient products may be made along the lines discussed, without in any way infringing on patented products.

Because of numerous factors affecting results, all the formulations cited herein, although believed to be reliable (most of them having emanated from the greatest laboratories in the world) should be tested thoroughly. It is especially recommended that manufacturers make their own tests to determine the suitability of the formulations, for the specific applications intended.

I am indebted to the manufacturers of the many chemicals named in the book, for information which I gathered either from direct correspondence or from their technical bulletins; also for their many samples with which I compounded some of the formulas. Special mention should be made of those companies that supplied photographs of equipment and installations depicting various processes.

I have tried to include all information pertinent to each formula, and credit has been given to all identified suppliers, so that additional information, if needed, may be obtained from the manufacturers of the basic materials used.

Finally, I am grateful to the publishers for valuable assistance in the organization and presentation of my quantities of material which now comes out as not merely a second edition of my *Oil, Fat, and Soap*, but as a much more comprehensive new book.

SEPTEMBER, 1966 BENJAMIN LEVITT

Contents

chapter 1

INTRODUCTION

Early History of Soap

The aphorism that "cleanliness is indeed next to godliness" preached by the English cleric, John Wesley, may truly have some basis in fact. It is thought by some that soap had its beginnings on the sacrificial altar of primitive peoples, when fat from slain animals dripped over the wood ashes, thereby combining with the potash. Although soap is mentioned in the Bible,[1] it is quite certain that the product bore slight resemblance to the products of the present day. It is claimed that the Phoenicians, who engaged in commerce on the Mediterranean about 600 B.C., were the first people to produce soap on a commercial scale.

Pliny [2] refers to it as an emollient and remedy for external use. He calls it *savinum caprinum cum calce*, and gives an account of soap being first made by the Gauls and being used as a cosmetic. It was made of goats' suet and beechwood ashes, and used in both solid and liquid forms.

Galen (131–200 A.D.), the Greek physician to the Roman emperor Marcus Aurelius, wrote that Gallic soap was best, because of its emollient effect and that it was useful for removing dirt from the body and clothes.

The Romans under the leadership of Julius Caesar, who later conquered Gaul, were quick to learn the Gallic method of soapmaking and introduced this useful cleansing agent for their personal comfort.

As early as the 7th century, soapmakers of Italy were organized as a craft guild. Under Charlemagne (768–814) soapmakers were recognized as craftsmen. Boissonade [3] wrote that in Venice, in the 15th century, bourgois capitalists speculated in the raw materials necessary to industry, and manufactured goods. He named among other commodities lard, potash, and soap. Italian and Flemish merchants sold tallow and lard daily at markets and at fairs. In the 14th and 15th centuries, soap was manufactured at Savona, Italy. The words *savon* (French), *jabon* (Spanish), *seife* (German), and *soap* are derivations from the name of that city.

Soon afterwards, Marseilles became a rival in the manufacture of soap. The natives had learned to make soap of olive oil and even perfumed it with floral odors, of which they had a plentiful supply. Marseilles soapmakers dominated the European market for four centuries.

Another center of manufacture was at Castile in Spain, which has given its name to a product made of olive oil. The name is still retained, but the composition of the soap has been modified to include other oils and fats.

American Developments

In colonial America, a "Mr. Brown, soapmaker" was granted admittance to Salem, Massachusetts in 1637. In 1716, John Lucena was granted the sole right to make Castile soap in the Rhode Island Colony. In 1735, *The Pennsylvania Gazette* carried an advertisement for "Super Crown Soap", describing its sweetness of flavor and immediate fine lather as rendering it pleasant for use by barbers.

Soap was also imported from Europe in colonial times. In the biography of Stephan Girard,[4] a colonial merchant of Philadelphia, a cargo from Marseilles is mentioned as comprising among other commodities, 500 cases of blue mottled soap.

About 1757, Bernard Gratz,[5] a Philadelphian, was in partnership with Michael Moses, a professional tallow chandler, for whom he supplied the capital. These kindred lines of business were often carried on under the same roof; for example, the partnership of Procter, a candle-maker, and Gamble, a soapmaker.

Benjamin Franklin, in his autobiography, mentions that his father who was a tallow chandler, tried to make a soapmaker of him, much to the disgust of the younger Franklin, who chose to become a printer instead.

Abraham Riviera, an uncle of Aaron Lopez, a leading businessman of the Revolutionary period, produced candles of sperm oil. Aaron's half brother, Moses Lopez obtained a license for the manufacture of potash.

Before and after the American Revolution, Solomon Simon and

his brother Samson were engaged in supplying the army. Samson became founder of the New York Chamber of Commerce. He was a partner in a spermaceti factory and one of the proprietors of the United Whaling Co.

J. L. Bishop,[5] stated that soap was manufactured in Boston in 1794. John Slidell & Co., at 50 Broadway in New York, was among the early soapmakers in the U.S.A. It was in this establishment that William Colgate learned the business. He started his own company on Dutch Street, in New York, in 1806. Colgate was the first soapmaker to render fats in his own plant.

In 1830 Jesse Oakley, of Newburgh, New York, introduced wrapped soap in cake form. Prior to that time, the grocer received soap in large blocks, from which he cut smaller bars and cakes for sale.

Philadelphia was established early as a center for soap making. Joseph Elkinton started his soap and candle factory in Philadelphia in 1831. Twenty-six years later, his son Thomas began making silicate of soda for use in the firm's soap. In 1861, silicate of soda was first offered for sale to others but the market for silicate expanded only slowly and it was not until 1904 that it was great enough to justify the discontinuance of soapmaking. This enterprise later became the Philadelphia Quartz Company.

The Glenn Perfumery and Toilet Soaps were known in Philadelphia in 1832. In 1848, Maas and Trebouillet obtained the first patent for the distillation of fatty acids, under vacuum.

Vroom and Fowler's *Walnut Oil Military Shaving Soap*, in tablet form, is said to have been invented in 1840. Thomson's Soap Foam, "the purest soap in concentrated powder form" is also an early product of Philadelphia. The labels for these soaps were reprinted recently in *The Philadelphia Inquirer* and may now be seen in the Atwater Kent Museum.

The Thomas Worsely Soap Works, in Combes Alley, Philadelphia,[6] was established in 1846 and sold in 1876 to Fels and Company, which had been founded in 1866. The latter became a part of the Purex Corp. in 1964.

In 1848, Alexander McConnell commenced the manufacture of fulling soap, tallow candles, and sal soda. J. Eavenson began to manufacture soap at 731 Hubble Street, Camden, N. J.

James Pyle and Sons were the originators of soap powder in the U.S.A. Prior to 1856, Pyle had sold washing soda crystals and bluing. In 1857, he started to make soap under the trade-mark "*O.K*" and a soap powder known as "Pearline." This business was later sold to Procter & Gamble.

The Los Angeles Soap Company is the oldest soapmaking firm west of the Mississippi and now the fifth largest in the U.S.A. It was started in 1860, in the Mexican village of Los Angeles.

B. T. Babbitt is credited with the production of pressed laundry soap in New York, in 1865.

The Pennsylvania Soap Works, of McKeone, Van Haagen and Company, established in 1854 on Callowhill Street, Philadelphia, was the largest soap plant in Pennsylvania. The use of rosin in soapmaking was begun shortly before the Civil War. A. Van Haagen developed a means of hardening rosin soap by using sal soda.

The recovery of glycerol from waste lye was an English invention which was improved in a U. S. patent granted in 1870. The world, however, is no longer dependent on the soap kettle for glycerol.

The first floating soap came into being when a workman stirred his batch of soap too long, thus aerating it sufficiently to make it buoyant. Ivory was its name.

The Industries of Philadelphia, published 1884, by Scharf and Wescott, lists thirty-one soap plants seven of which made perfumed soap. At that time the Dallet family of three generations, were makers of common and fancy soap and of moulded and dipped candles.

Besides the names listed, there were such names as Enoch Morgan and Sons, Hay, Brown, J. C. Hull's Sons, and Fay, all of whom started in the early years of our country. As in many other industries, there has been a great number of mergers, so that most of the early companies have lost their identity.

Scientific Study of Oils, Soap, and Lye

The first scientific study of soap was made in 1741 by Geoffroy, a French chemist, who pointed out that the fat recovered from soap by treatment with an acid was not of the same composition as the original fat. He showed that the fat derived from the soap dissolved in alcohol, whereas the original fat was insoluble in alcohol. It is to be noted that with the exception of castor oil (*ricinolein*), none of the neutral fats or oils is soluble in alcohol.

Scheele (Swedish chemist) in 1783, while boiling lead oxide with olive oil in making lead plaster, discovered a sweet substance as a by-product, which he called *Oelsuss* (sweet oil), or *principium dulce oleorum* (sweet principle of oils). The substance is the glycerol of today. Chevreul (another French chemist), in 1815, based his work on these discoveries. He disclosed the true nature of oils, glycerol, and soap. He showed that soap is the result of the interaction of the acids in the fats, with an alkali.

Up to the time of the French revolution, the lye necessary for soapmaking was obtained by lixiviation of wood ashes or the ashes from kelp. It is worth noting that the first patent granted by the U. S. Patent Office—to Samuel Hopkins on July 31, 1790—covers an apparatus and process for making pot-ash and pearl ash. This historical document, written on sheepskin, was signed by President George Washington, in New York, which was then the national capital.

During the French revolution, when the supply of kelp was cut off by an embargo, a prize was offered by the French Academy of Science for a method of manufacturing soda ash from salt. LeBlanc won the prize in 1790 by using sulfuric acid, limestone, and coal, in the following three-step process:

1. Common salt is warmed with sulfuric acid, forming sodium sulfate (salt cake) and hydrochloric acid:

$$2\,NaCl + H_2SO_4 = Na_2SO_4 + 2\,HCl$$

2. Sodium sulfate is heated with coal and limestone:

$$Na_2SO_4 + 2\,C = Na_2S + 2\,CO_2;$$
$$Na_2S + CaCO_3 = Na_2CO_3 + CaS$$

3. Black ash is lixiviated with water to dissolve out, and then crystallize, the sodium carbonate.

Although this process was used for many years, it was cumbersome and it produced impure soda ash. It was superseded in 1863 by the Solvay process, which is based on the reaction of salt with ammonium bicarbonate. Ammonia and carbon dioxide are led alternately into a cold concentrated solution of salt under pressure:

$$NaCl + (NH_4)HCO_3 = NaHCO_3 + NH_4Cl$$

The sodium bicarbonate settles out of the cold concentrated ammonium chloride solution. It is then calcined to yield soda ash and carbon dioxide.

Caustic soda was produced by causticizing soda ash with lime. At present, practically all caustic soda is produced by the electrolysis of brine.

Synthetic alkali was first used in soapmaking in 1823. It was manufactured in England by James Muspratt by means of the LeBlanc

TABLE 1:1

PRODUCTION AND CONSUMPTION OF OILS, FATS, AND DERIVATIVES*
(Millions of Pounds)

	1961	*1963*	*1964*
Baking and Frying Fats	2,456.1	2,584.3	2,664.1
Butter	1,536.0	1,453.0	1,468.0
Castor, consumption	115.7	124.5	136.6
Coconut, crude	498.0	348.3	327.6
Coconut, refined	463.0	554.5	506.0
Cooking and Salad Oils	2,123.7		
Corn, crude	335.8	390.5	413.9
Corn, refined	321.7	363.3	393.1
Cottonseed, crude	1,793.8	1,917.0	1,932.4
Cottonseed, refined	1,506.0	1,577.3	1,600.0
Fatty Acids	706.0	835.6	936.0
Fish, including Mammal oil, except Sperm	258.1	185.8	180.2
Glycerol, crude	270.2	302.1	328.1
Crude used in refining	269.1	295.6	313.2
Imports	18.4	2.9	11.8
Refined	269.4	303.2	320.1
Lard	2,384.0	2,373.0	2,388.0
Linseed, raw and boiled	426.3	399.1	443.6
Margarine	1,723.7	1,793.6	1,857.4
Olive Oil, sulfured imports		0.7	0.2
Olive Oil, edible		32.9	66.9
Palm Kernel	53.6	52.3	63.8
Palm Oil, consumption	65.8	44.2	37.5
Peanut, crude	96.8	99.1	123.3
Peanut, refined	94.6	79.2	60.5
Safflower, crude		116.1	withheld
Safflower, refined		76.9	75.4
Sperm, consumption			
Selected products	46.6	36.1	33.6
Imports		65.9	58.9
Soybean, crude	4,442.3	5,053.2	4,943.8
Soybean, refined	3,592.7	4,033.7	4,591.8
Edible Tallow	443.7	527.9	553.2
Inedible Tallow and Grease			
Production	3,554.3	4,156.5	4,565.7
Consumption: Soap		684.0	
Fatty Acids		448.8	
Feeds		754.5	
Lubricants, etc.		80.2	
Other products		166.3	
Tall, crude	826.3	991.3	1,104.0
Tall, refined	139.2	150.4	143.6
Tung, production crude	16.0	11.8	24.5
Consumption, selected products	37.8		
Vegetable Foots	278.9	284.0	261.1

* From U. S. Dept. of Commerce.

process. Commercial production of caustic soda dates from 1854, when William Gossage, in England, produced an improved product for the use of soapmakers and papermakers. Although Solvay took out the first patent in Belgium in 1863, it was not until 1881 that a plant for the Solvay process was established at Syracuse, New York.

Important contributions to the study of oils were made by such men as Mercer (sulfonated olive oil), Twitchell (fat-splitting), Wesson (refining and deodorizing), Sabatier and Normann (hydrogenation), Lewkowitsch (technology), and many others. These authors and their work will be discussed in other chapters under the appropriate headings.

CLASSIFICATION OF FATS AND OILS

Fats and oils are complex mixtures of glycerides. They are esters of fatty acids and the trihydric alcohol, glycerol.

Cottonseed oil, for example, consists of the glyceryl esters of linoleic, oleic, palmitic, myristic, and stearic acids, mentioned here in order of their decreasing percentages.

The distinction between fats and oils is a purely arbitrary one, based on their physical state at ordinary temperatures, the oils being liquid and the fats, pasty or solid. There are, however, many exceptions, such as coconut, babassu, palm, and palm kernel oils, which range from pasty solids to hard waxlike solids in a temperate climate. The latter have received the designation "oil" only by convention.

Another classification of oils is based on their origin, such as animal (terrestrial or marine), vegetable, and mineral. As the members of the last class consist almost entirely of hydrocarbons and have their origin in petroleum, the scope of this book does not permit discussion of mineral oils.

The number of products available as aerosols seems to be unlimited and increasing. Many practical suggestions for this form of application will be discussed in Chapter 1 of Volume 2.

REFERENCES

1. *Old Testament*, Jeremiah, II:2; Malachi III:2.
2. Pliny, *Historia Naturalis*, Book XIII.
3. P. Boissonade (Translated by E. Power), *Life and Work in Medieval Europe*, p. 103; Alfred Knopf, New York, 1927.
4. H. E. Wildes, *Lonely Midas;* Farrar and Rinehart, New York, 1943.
5. J. L. Bishop, *History of American Manufactures*, 1866.
6. J. Thomas and Thomson Wescott, *The Industries of Philadelphia;* L. H. Eberts, Philadelphia, 1884.

chapter 2

ANIMAL FATS AND OILS

The most important fats and oils of animal origin are tallow, lard, whale oil, and cod liver oil. Degradation products of animal fats (possibly mixed with vegetable oils and fats) house grease, and garbage grease may also be placed in this category. Wool grease, derived from washings of raw wool, and sod oil, which is recovered from the chamoising process of sheep skins, are also members of this class.

TERRESTRIAL ANIMALS

Edible Tallow

Among the animal fats, tallow is the most important. The grade of tallow depends on the quality of the raw material from which it is rendered. Edible tallow is obtained by rendering clean suet from wholesome carcasses of cattle in the packing house. *Oleo oil* is pressed from this grade of tallow and used to form the basis of oleomargarine.* The stearin, which is the harder (saturated) portion of the fat, remains in the press. This is used in the manufacture of shortenings.

Inedible Tallow

Inedible tallow is rendered from suet and butchers' trimmings that are unsuitable for food. The product is graded according to color and free fatty acids. It may be refined and sold as *acidless tallow oil* for compounding with mineral oil to form cylinder lubricants and for other purposes.

The suet and butchers' trimmings are loaded into a *rendering tank* (Fig. 2:1), which is a steam-jacketed horizontal tank, equipped with

* Since the polyunsaturates have come to the attention of the public, less oleo oil is used. At least its inclusion in oleomargarine is not being publicized.

FIG. 2:1. HORIZONTAL MELTER FOR FATS
(*French Oil Mill Machinery Co., Piqua, Ohio*)

FIG. 2:2. HYDRAULIC CAGE PRESS
(*French Oil Mill Machinery Co., Piqua, Ohio*)

a stirring device. The charge is cooked under pressure for the first hour, so as to disintegrate the cellular tissue and to free the fat. Vacuum is then applied, while cooking is continued for several hours. This drives off the moisture. The fat is then drained into settling tanks and the "tankage" cracklings is squeezed in a hydraulic press (Fig. 2:2) to free it from the remaining tallow, after which it may be ground for fertilizer or sold as meat scraps for chicken feed. It consists of tissue that is rich in nitrogen and phosphates and contains three percent or less of fat. (Sometimes the residual fat is extracted with volatile solvents.)

In some of the newer rendering processes, the stock, including bones, is ground before it goes into the rendering tank. The rendering time thus is shortened. In the old method of rendering, known as the *wet process*, the moisture that resulted from the disintegration of the fatty tissues remained in the cooker. No vacuum drying was employed. The water had to be evaporated from the glue stock. This "stick" was also used in the manufacture of fertilizer.

Acidless Tallow Oil

To produce this acidless oil, prime tallow is melted and then held at 85–90°F for several days to grain thoroughly. It is then pressed. For refining, the batch is stirred with the requisite amount of caustic soda solution and salt is added to facilitate settling of the soap stock. The clear oil is then pumped off. Tallow oil is used for compounding cylinder oils. The stearin is used for heavy-duty lubricants.

The specifications of General Motors Corporation for steam cylinder oil are as follows:

> The oil shall be a homogeneous mixture of acidless tallow oil and properly refined mineral cylinder stock, free from dirt, soluble acid, or extraneous matter. The acidless tallow oil content ranges with the different types of cylinder oils, from 5 to 8 per cent with a maximum of 0.2 per cent of free fatty acid in the finished oil.

Lard

Lard is a soft fat derived from hogs. The best grade of so-called *leaf lard* is a pure white product, obtained from the leaf fat that surrounds the kidneys and intestines of the hog. It was formerly boiled in open kettles at low temperature.

Lard from other parts of the animal is rendered by direct application of steam. The process has now been automated, using the French

Drip Rendered Lard Cooker, shown in Figure 2:3. This combines the operation of rendering and refining in a one-stage process, and makes it possible to produce a uniform high-grade lard from any kind of pork fats.

3,388 million pounds of lard were produced in 1964 in the United States.

Because lard is a soft fat and does not have good keeping qualities, it is often improved by hydrogenation and by the addition of anti-oxidants.

FIG. 2:3. DRIP-RENDERED LARD COOKER
(French Oil Mill Machinery Co., Piqua, Ohio)

TABLE 2:1

CHEMICAL AND PHYSICAL TESTS ON PACKING-HOUSE PRODUCTS

Stock	*FFA* %	*Color*	*Iodine No.*	*Sapon No.*	*CT* °*F*	*Pour* °*F*	*SUV* @ *100°F*	*Flash* °*F*	*Fire* °*F*	*Titer* °*C*
Prime Lard Oil	0.5	1½ NPA	65–75	190–195		40	200–220	550	600	32–35
Ex. or O.P.	5.0	3 NPA	65–75	190–195		45	200–220	500	550	32–36
No. 1 Lard Oil	20.0	6 NPA	65–80	190 min		45	230 max	450 min	1500	32–36
Neatsfoot Oil	1.0	1¼ NPA	70–78	same	20		210	550 min	600	22–26
Neatsfoot Oil	1.0	1¼ NPA	65–75	,,	30		210	550	600	26–30
Neatsfoot Stock	1.0	11 FAC	65–72	,,						26–30
Acidless Tallow Oil	1.0	2 NPA	50–60	192 min		95	55 @ 210	550	600	38–42
Choice White Grease	5.0	7 FAC	68	192						37–40
Yellow Grease	15.0	37 FAC max								36
House Grease	20.0	39 FAC	65–70	195						39
Edible Tallow	1.0	5 ,,								41.5
Fancy ,,	4.0	7 ,,								41.5
Choice ,,	5.0	9 ,,								41.0
Ex ,,	6.0	13.11B								40.5
Sp ,,	12.0	19.11C								40.5
No. 1 ,,	15.0	33								40.5

Abbreviations:

CT	Cold Test	FFA	Free Fatty Acids	OP	Off Prime
Ex	Extra	max	Maximum	Sapon	Saponification
FAC	Fat Analysis Committee, American Chemical Society	min	Minimum	Sp	Special
		NPA	National Petroleum Association	SUV	Saybolt Universal Viscosity

Although the largest portion of lard is consumed as such, a good deal is used in the preparation of shortenings, which contain in addition, other animal fats and/or vegetable oils and stearin.

Lard Oil

Prime lard oil is the fluid expressed from white grease, an inedible lard, which is produced by rendering the fat of condemned hogs.

The oil is obtained by seeding or graining the grease which is kept at 50°F in trucks of about 800-lb capacity. By *seeding* is meant the crystallizing and separating of the stearin. Packages or *puds*, as they are called in the packing house, are made up by wrapping several pounds of the seeded grease in pieces of six-ounce duck about twenty inches square. The puds are placed on steel plates, layer on layer, and then pressed in a hydraulic press. This method of manual treatment has been mostly superseded by extraction with solvents.

The best grade of lard oil is prime Winter Edible, produced from edible lard. It is used in baked goods, confectionery, margarine, and other food products.

Technical lard oils are graded as follows:

Prime Winter Inedible,
Off Prime,
Extra No. 1, and
Extra No. 2.

Prime Winter Inedible and *Off Prime* grades are produced from white grease (see *Miscellaneous Greases*).

The darker grades of lard oil are produced from house grease. This is given a preliminary purification by means of sulfuric acid to precipitate the albuminous matter and other impurities.

The characteristics of each of the technical grades are listed in Table 2:1.

Lard oil is used in the manufacture of soaps, lubricants, and metal-cutting compounds, in signal lamps, and in textile oils.

In the treatment of textiles, lard oil is used chiefly for oiling wool.

Miscellaneous Greases

White Grease:

White grease is a light-colored fat, produced by rendering the fatty offal that is unfit for use as lard and also obtained from condemned carcasses. It is usually low in free fatty acids and has a titer of about 34°C. Its chief use is in the pressing of the better grades of lard oil.

House Grease:

House grease is produced from kitchen drippings collected from hotels and restaurants. It is brown in color, has a free fatty acid content of 12% or more, and the titer is about 37°C. It is a mixture of animal and vegetable fats and oils. It is used in making yellow soaps and for pressing lower grades of lard oil.

TABLE 2:2

STANDARD GRADES, SPECIFICATIONS AND QUALITY TOLERANCES FOR DOMESTIC TALLOWS AND GREASES [a]

Rule 5. The standard grades of tallow and grease as set forth below, are the official New York Produce Exchange domestic grades. The specifications therefore shall govern domestic trading in tallow and grease under these rules, unless the written contract specifically provides otherwise.

Grade	*Specifications*			
	Min ° Titre	Max % F.F.A.	Max F.A.C.	MIU % [b]
TALLOW				
Extra Fancy No. 1	43.0	2.3 *	5	1
,, ,, No. 2	42.5	2.3 *	5	1
,, ,, No. 3	42.0	2.3 *	5	1
** Bleachable Extra Fancy	43.0	3.0	1.0 R&B	1
Industrial Extra Fancy	42.5	3.0	5	1
** Bleachable Indust. Ex. Fancy	42.5	3.0	1.0 R&B	1
Fancy	41.5	4.0	7	1
** Bleachable Fancy	41.5	4.0	1.0 R&B	1
,,	41.5	4.0	1.5 R&B	1
,,	41.5	4.0	2.0 R&B	1
Choice	41.0	5.0	9	1
Prime	40.5	6.0	13–11B	1
Special	40.5	10.0	19–11C	1
No. 1	40.5	15.0	33	2
No. 3	40.5	35.0	none	2
No. 3	40.5	35.0	none	2
No. 2	40.0	20.0	37	2
GREASE				
Choice White Grease (all hog)	37.0	4.0	13–11B	1.0
A White	37.0	8.0	15	1.0
B White	36.0	10.0	19–11C	1.0
A Yellow Grease	36.0	15.0	37	2.0
B Yellow ,,	39.0	15.0	39	2.0
House ,,	37.5	20.0	39	2.0
Brown ,,	38.0	50.0	none	2.0

* To be specified in contract.

** Refined and bleached color specification: no F.A.C. color requirement.

a From National Renderers' Association, Des Plaines, Ill.; 4–20–65.

b All basis; no reciprocal allowance to seller.

Garbage Grease:

Garbage grease is produced by dehydrating municipal garbage and then extracting it with a solvent. The solvent is distilled off and the grease is recovered. It is a dark, vile-smelling fat which is too impure even for soapmaking. The free fatty acids in such grease may amount to 60%, which indicates the extent of the decomposition that has taken place.

To prepare this substance for soapmaking it may be converted into fatty acids by means of the Twitchell process. The fatty acids are then distilled and pressed to yield stearic acid and red oil (oleic acid).

Chemical and physical constants for common packing-house products are shown in Table 2:1; and specifications for tallows and greases, in Table 2:2.

Neatsfoot Stock and Oil

Neatsfoot oil is obtained from the shin bones and feet of cattle. The feet are cut off, washed, and trimmed. The hoofs are removed, the bones are sawed into small pieces, and then boiled with water in an open kettle to separate the fat. From five to ten hours of boiling are required. The fat is skimmed off, strained into a steam-jacketed kettle, and heated to 240°F. Finally the oil is filtered through flannel bags.

The stock, as received at the refinery or pressing plant, is seeded out at 34°F for about two weeks and then pressed. The first yield of neatsfoot oil, called *Export Pure*, constitutes the finest grade, having a cloud point * of 16–18°F. This oil is used for finishing fine leathers and oiling delicate machinery.

The stearin remaining in the press is remelted and again allowed to seed out at 40°F, to produce neatsfoot oil of 30°F cold test.** This grade of oil is used by throwsters in silk-soaking emulsions. Combined with light mineral oil, it is used as a basis for knitting oils, and as a stainless oil for hosiery knitting needles, because it can readily be washed out. It is also used in the fat-liquoring of leather.

FROM MARINE ANIMALS

Whale Oil

Whaling is now practiced in the South Atlantic and Antarctic Oceans. Some years ago there was some whaling also in the Pacific

* The cloud point of an oil is the temperature at which a permanent cloud forms under standard conditions of chilling the oil.

** The cold test of an oil indicates the lowest temperature at which it will just flow. Below this temperature the oil begins to solidify and cannot be used as a lubricant.

Ocean, but the whale population has been almost entirely wiped out there. This fact necessitated governmental regulations and international agreements on whaling operations.[1]

As the whale—not a cold-blooded fish, but the largest known mammal—comes to the surface for breathing, it is harpooned by a missile weighing 120 pounds, shot from a harpoon gun. A delayed-action bomb in the head of the harpoon explodes deep inside the whale, inflating the carcass and thus keeping it afloat. The whale catchers are small tug-like ships, about 100 feet long, built to turn quickly. These boats haul their catch to larger factory ships, which are equipped to cut up and render the blubber and to extract sperm oil from the "case" in the animal's head. The blubber which is the layer of fat immediately under the skin, is cut off and rendered in either open or pressure kettles. The skin and bones are chopped up for fertilizer. According to reports, one ship can process 1600 whales and produce 15,000 tons of oil on one voyage.

The oil received from the kettle is refined and pressed for proper cold test.

Oil produced immediately after the whale is caught, may be used in edible products. Considerable quantities of such oil are consumed in Europe for the production of margarine.

According to Norwegian Specification No. 487, dated 1939 there are four grades of whale oil, based on the content of free fatty acids, and on color:

No. 1. Pale straw color not exceeding 2% of free fatty acids; color 3 red and 35 yellow Lovibond, in a 40-mm cell.
No. 2. Amber Yellow and not exceeding 6% of free fatty acids; color 10 red.
No. 3. Pale brown and in excess of 15% of free fatty acids; color 12 red.
No. 4. Dark brown and in excess of 15% of free fatty acids; color darker than No. 3.

Whale oil is to a large extent hydrogenated, to produce a white fat somewhat like tallow. When so treated, the oil loses much of its fishy odor and is improved in color. It can be used as it is, or it may be split by acid, to produce stearic acid and glycerol. The oil is used in the metal trades for manufacturing cutting oils, and for burnishing compounds. Another application is as an adhesive in agricultural sprays. It is also used as an extender for linseed oil in paints, varnishes, and printing ink, and as a lubricant.

Fish Oils

The fish oils of commerce may be divided into two classes: those that are produced from the whole fish or offal from fish canneries; and those obtained from livers. Among the whole fish utilized for oil production, menhaden, sardine, herring, pilchard, and salmon are the most important.

Herring Oil:

A small quantity of herring oil is produced in the U.S.A. as a by-product of canneries. In Norway, Sweden, Iceland, and Alaska, this oil is obtained from the whole fish. As some of the fish are stored to overcome seasonal work, the oil reaches the market varying in color from pale yellow to dark brown. The process for manufacturing it is similar to that by which other fish oils are produced.

Menhaden Oil:

Menhaden (*Brevoortia tyrranus*) is related to the herring, which it resembles. It is caught on the Atlantic seaboard. It is used solely for conversion into oil and fish meal. The oil is produced in cookers 2.5 feet in diameter and 40 feet long. The fish are fed into the cooker at one end and are propelled by means of revolving blades. Live steam admitted to the cooker disintegrates the fish in fifteen minutes. The mass is continuously squeezed in a screw press, the oil mixed with water, and the suspended solids is passed into settling tanks. The oil is skimmed off twice and the dregs are further heated to effect complete separation of the oil. The residual solids are dried and sold as feed.

Pilchard Oil is produced on the west coast of Canada by the method used for sardine oil.

Salmon Oil:

Salmon oil is produced entirely from the offal from canneries on the Pacific coast of the U.S.A. and Canada. The offal amounts to 33% of the total weight of the fish caught, and includes heads, tails, and viscera with livers. This refuse is cooked to recover the oil, and the meal is dried for feed. The oil consists of 15% of saturated and 85% of unsaturated glycerides.

Sardine Oil:

Most of the sardine oil of commerce comes from the west coast of the U.S.A., and from Japan. The sardines are cooked, dried, and extracted with petroleum solvent to yield oil and meal. By an alternate method, the sardines are handled somewhat in the manner of menhaden; that is, cooked and the oil allowed to separate out. Then, the whole slurry is centrifuged.

Applications

Some fish oils are used for low-grade soap, but for the better grades, the oil must be hydrogenated. Mixed with linseed oil, fish oils are used in paints. By the newer processes of solvent fractionation of unsaturated constituents, fractions can be isolated that have good drying properties, so that they are very useful extenders for drying oils.

Kettle-bodied and blown fish oils, which have been oxidized and made more viscous, are used in paint, linoleum, and printing ink, and in the manufacture of patent leather. Considerable quantities of fish oil are sulfonated and used for fat-liquoring emulsions in the manufacture of leather.

Fish Liver Oils

In therapeutic oils, the maximum content of free fatty acids should be 0.5 percent. The fresh livers are cleaned to remove gall bladders. They are then washed and cooked for 30 minutes, in corrosion-proof, open kettles. The cooking separates the oil and decomposes all fat-splitting enzymes. The oil that floats on the surface is drawn off and allowed to settle. It is next centrifuged and filtered to remove water and extraneous matter. It may be bleached with fuller's earth, and further refined by chilling out the stearin at 15°F and filtering. The residue remaining in the kettle is pressed to yield lower grades of vitamin-containing oil, which is used as feed for domestic animals and poultry.

Some liver oils—from halibut, shark, swordfish, and tuna—have been reported to contain 100,000 to 300,000 U.S.P. units of vitamin A per gram, and tuna-liver may contain up to 70,000 units of vitamin D.

Halibut-liver oil is sold with a potency of 60,000 to 65,000 units of vitamin A, and 10,000 to 13,000 units of vitamin D per gram of oil.

The vitamin D content of the oils varies and is independent of the vitamin A activity. Molecular distillation is one of the methods used to separate and concentrate vitamins.[2]

Cod Liver Oil

Official cod-liver oil (*Oleum Morrhuae;* U.S.P. XVI) is the partially destearinated fixed oil obtained from fresh livers of *Gadus morrhua* and other species of the family *Gadidae.*

Cod liver oil contains in each gram not less than 255 micrograms (850 U.S.P. Units) of vitamin A and not less than 2.125 micrograms (85 U.S.P. Units) of vitamin D.

Cod Liver Oil may be flavored by the addition of not more than 1% of a suitable flavoring substance or a mixture of such substances.

One gram of vitamin A contains 4,000,000 international U.S.P. units.[3]

Recommended daily dietary allowances (revised 1958)[4] are

	Internat. Units Vitamin A	Internat. Units Vitamin D
for Children 1–3 yrs.	2000	400
for Children 16–19 yrs.	5000	400

Cod Oil:

Cod Liver Oil that is unfit for therapeutic purposes is sold as *cod oil.* Its principal use is in the chamoising of sheep skins. They are tanned by oxidizing the oil on the skins, while drying in hot rooms.

Cod oil is sulfonated and used in the fat-liquoring of leather. It is also used in the manufacture of low-grade soap for agricultural purposes.

BY-PRODUCTS AND DERIVATIVES

Butter

Butter is not a true fat. It is an aqueous emulsion of the fat contained in cow's milk, together with dissolved casein and other milk solids and salt. It is a water-in-oil emulsion of the following approximate composition:

Fat (average)	81.10000%
Water	15.5
Salt	2.3
Protein (Casein)	0.6
Lactose	0.4
Riboflavin	0.00001
Calcium	0.02
Phosphorus	0.01
Diacetyl	0.0003
Acetoin	0.0015
Niacin	0.0001

Vitamin A, 7390 I.U.; Vitamin D, 150 units/lb.

Oleomargarine

Oleomargarine has become one of the important items of food. In fact, since 1952 when production was 1,271,000,000 pounds, consumption of it has exceeded the consumption of butter. In 1961, 1,484,000,000 pounds of butter were produced but the production of margarine was 1,724,000,000 pounds.

Oleomargarine, or *margarine*, was invented at the instigation of Napoleon III, during the Franco-Prussian war. He offered a prize for the best substitute for butter. A French chemist, Mege-Mouries, was awarded the prize for his invention. It consisted of beef tallow, rendered at low temperature. This is then chilled to separate out the stearin, while the softer portion of the fat was churned with milk and the tissues from the udder of a cow, to impart the odor and taste of real butter.

Since that time, many improvements have been made in the process. Additions of egg yolk were once made to facilitate emulsification of the ripened milk with the fat. Later, egg yolk was replaced with lecithin, isolated from soybean oil. Other emulsifiers used are mono and diglycerides, which prevent the finished product from "leaking" or water separation. Mono- and distearin sulfoacetates are used to prevent spattering.

The method followed at present is discussed in Chapter 6.

Lanolin

Sheep's wool, as it is received at the scouring plant, contains 40–70% of impurities, a substantial portion of which is a grease that exudes from the sebaceous glands in the skin of the animal to protect and soften the fleece. This grease is removed either by washing the raw wool with soap and soda ash, or by extraction with a volatile solvent. When the extraction method is employed, the solvent is distilled off, leaving the wool grease which is either sold "as is," or purified and sold as *lanolin*.

Lanolin * is sold in two forms. The *anhydrous* product, containing not more than traces of water; and the *hydrous* lanolin, which contains 25–30% of water. It is a yellowish, tenacious, semi-solid, greasy material, with a weak, characteristic odor. It melts at 38–40°C. It is insoluble in water, but miscible with about twice its weight of water, without separation; it is sparingly soluble in cold ethanol and more in hot ethanol. It is also soluble in benzene, chloroform, carbon disulfide, acetone, and ligroin.

Lanolin is used as a base for ointments and creams. It is claimed that lanolin is more rapidly absorbed by the skin than some other

* Suppliers: 38, 83, 85, 128.

oils and ointment bases. It is employed in soap as a superfatting agent, to counteract alkalinity and to serve as an emollient for the skin.

The major constituents of lanolin are cholesterol, iso-cholesterol, together with their esters and fatty acids. Some of these components have been isolated, and others have been converted into useful related products.

Modified Lanolin

To take advantage of all the good properties of lanolin and make it easier to handle, manufacturers have invested much research in the development of modified products.

These now include natural liquid lanolin, and oil-soluble, water-soluble, and alcohol-soluble products, which contain various esters, cholesterol, mineral oil, and other substances suitable for the intended purpose.

One of these liquid lanolins, Viscolan (American cholesterol), is an unaltered natural liquid ester fraction of lanolin. It is soluble in oils, propellents, silicone fluids, and other materials for formulations. It is also used in creams, lotions, and pharmaceutical preparations.

A solid ester fraction (Waxolan) is a plastic natural wax. It is used in anhydrous and emulsified hair dressings and other toilet preparations.

Some other oil-soluble lanolin products are:

- Fluilan (Croda): Liquid lanolin esters.
- Isopropylan (Robinson-Wagner): Contains 33–50% of unreacted lanolin, with the balance esters.
- Lanogene (Robinson-Wagner): Liquid fraction containing 15–17% of cholesterol.
- Lantrol (Malmstrom): A liquid fraction consisting of esters of lower molecular weight, which are completely soluble in mineral oil.
- Liquilan (Robinson-Wagner): Liquid lanolin esters soluble in mineral oil, can be used in waterless hand soap.

Some of the alcohol-soluble lanolin products are:

- G 1441, 1471, and 1790 (Atlas) are polyoxysorbitol lanolin derivatives.
- Ethylan (Robinson-Wagner) oily liquid containing approximately 35% of pure untreated lanolin, dissolved in a lanolin

derivative. It is also soluble in fatty acid esters, mineral oil, vegetable oils, and aerosol propellents.

A few of the water-soluble lanolin products are:

G 1441, 1471, 1790 (Atlas)

Lanogel series (Robinson-Wagner): Polyoxyethylene lanolin

Nimcolans (Malmstrom): Similar to preceding.

Absorption Bases

Absorption bases, valuable as carriers for special ingredients of pharmaceutical and cosmetic ointments and creams, are mixtures of lanolin and suitable components.

Some of the well-known Lanolin Absorption Bases are:

Alcolans (Robinson-Wagner): Emollient water in oil emulsifiers containing from 9–70% of lanolin fractions, with the balance mineral oil.

Eucerin (Duke): A W/O emulsion

Falba (Pfaltz-Bauer), Isco (M N), Isocreme (C), Isolan (G), Nimco (M), Sterolan (R W): A sterol absorption base, oil soluble, water-dispersible, liquid non-ionic surfactant with emollient action.

Hychols (Croda): A series of nonionic surfactants in which the hydrocarbon portion of the molecule is due to mixed sterols (cholesterol) trimethyl sterols (lanosterol) and aliphatic alcohols of high molecular weight: an oil-in-water emulsifier.

All applications of these commercial products are discussed in Volume 2.

Dégras

Dégras is crude wool grease obtained by washing or extracting wool with volatile solvents. It is a dark brown, solid, amorphous grease with a disagreeable odor. It has the following physical and chemical constants:

Specific gravity	0.9322–0.9449
Solidification point	38–40°C
Saponification number	84–127
Iodine number	15–21.5
Acid number	0.5–4.3

It is used for the manufacture of lanolin, leather-stuffing and leather-dressing compounds, soaps and drawing compounds, printing inks, steam cylinder oils, rust preventives, rope dressing, metal cutting, and many other uses.

Moellon Dégras:

Moellon Dégras,* sometimes called *sod oil,* is a thick brown oil which is pressed from chamois skins, after they have been tanned. It is of variable composition as it contains a considerable amount of unsaponifiable, chiefly cholesterol and oxidized fish oil. It is used in retanning leathers used for belting. It lubricates the fibers and moisture-proofs them, so that the leather becomes very supple. It is also used in lubricants.

Sperm Oil

Sperm oil should really not be classed as an oil, but as a liquid wax. It occurs in the head of the whale, in admixture with a more solid wax. By chilling and pressing, the fluid is separated out.

Sperm oil is a thin yellowish oil with a fishy odor. It is used as an illuminant, as a quenching oil in tempering steel, in soaps, compositions for dressing and finishing leather, in lubricants and automatic transmissions. Its saponification number is 123–147; its iodine number, 80–84. It is soluble in chloroform and ether. There are several grades: *Crude*, *Natural*, and *Bleached Winter*.

Spermaceti

Spermaceti, or *Cetaceum* (U.S.P. XVI) is not a fat, but a waxy substance extracted from sperm oil obtained from the head of the sperm whale, *Physeter macrocephalus, Linne* * (Fam. *Physeteridae*). It is a white crystalline solid wax, which is used in the production of cetyl alcohol and as a base for ointments and cosmetic creams. It consists chiefly of cetyl palmitate, free cetyl alcohol, esters of lauric, stearic, and myristic acids, and esters of oleyl alcohol.

Spermaceti is insoluble in water, nearly insoluble in cold ethanol, and only slightly soluble in cold solvent hexane. It is soluble in boiling ethanol, in ether, chloroform, carbon disulfide, and in fixed and volatile oils. It contains about 50% saponifiable matter and it has the following constants:†

Melting Point	42–50°C
Saponification No.	120–136†
Iodine Value	3–4.4

---

* Suppliers: 38, 83, 85, 129.
† Suppliers: 13, 149, 168.

Sterols

The sterols are crystalline hydroaromatic alcohols. They occur principally in the unsaponifiable portion of fixed oils and fats, from 7 to 8% in halibut-liver oil, down to as low as 0.08% in beef tallow.

The sterols of animal origin are classed as *zoosterols*, those from plants as *phytosterols*, and those originating from fungi, as *mycosterols*. Although they are of minor importance as ingredients of all oils, the sterols are of interest because some of them serve as the starting point for the synthesis of vitamin D and of sex hormones.

Testosterone has been synthesized from cholesterol and desoxycorticosterone from stigmasterol; see Chapter 2.

The sterols serve as a means of identifying the oils from which they are derived. The acetates are employed for this purpose. Cholesterol acetate melts at 114°C, and phytosterol acetate has a melting point of 148–149°C. The presence of cholesterol indicates animal oil, while phytosterol indicates vegetable oil.

Cholesterol

The principal sterol of animal origin is *cholesterol*, derived solely from lanolin. It assists the body to use fat, and is used also in the body for making bile salts, hormones, and vitamin D. A dehydrogenated cholesterol, 7-dehydrocholesterin, on irradiation with ultraviolet light, produces vitamin D_3, which is identical with that of fish-liver oils.

Cholesterol U.S.P. is a purified crystalline powder, recommended as a surface-active agent and moisturizing emollient for preparations used in the treatment of skin, hair, and scalp, for hypoallergenic cosmetics, protective water-absorbable ointment bases, and both water-in-oil and oil-in-water emulsions. It melts at 147–149°C.

Pure cholesterol is claimed to be nonsensitizing even when tested on those individuals allergic to wool fat. (More on cholesterol under *Safflower oil, Chapter 3.*) *

* Supplier: 4.

REFERENCES

1. D. F. Tressler, *Marine Products of Commerce;* Chemical Catalog Co., New York, 1923.
2. U. S. Patent 1 925 559. K. C. D. Hickman (to Eastman Kodak), 9/5/33.

3. A. L. Bailey, *Industrial Oil and Fat Products;* Interscience Publishers, New York, 1945.
4. *Food Yearbook;* U. S. Department of Agriculture, Washington, D.C., 1959.

chapter 3

VEGETABLE FATS AND OILS

With the exception of cocoa butter, none of the vegetable fats has become commercially important in the United States. The familiar oils remain, but new applications have been found for some of them, and their ranks have been joined by safflower, soybean, and others, which are increasingly important.

Castor Oil

Castor oil is a clear, colorless or yellowish oil, extracted from the beans of *Ricinus Communis*. The principal sources of supply are Brazil, India, Manchuria and Russia. Its specific gravity is 0.96–0.97—higher than that of any other untreated vegetable oil. It is soluble in the common organic solvents; (for other properties, see Table 3:2).

The decorticated kernels are pressed cold. The oil so obtained is used medicinally. The oil obtained by a second pressing, is used for commercial purposes. It may also be obtained by extraction with a volatile solvent. The yield of oil is about 40 percent.

Castor oil is refined by steaming. This treatment coagulates the dissolved albumin which can be removed by filtration. The press-cake is unsuitable for cattle feed because it contains toxic albumins, ricine and ricinine. Both of these can be removed according to a German patent,[1] by extraction with hot water. When so treated, the residue may be used in fodder, however, it is used mostly for fertilizer.

In addition to its medicinal use, castor oil has many industrial applications. It is used as a plasticizer in patent leather. In its sulfonated form, it is marketed as Turkey-Red Oil and used in the dyeing and finishing of textiles and leather.

About half the castor oil produced is consumed by the paint and varnish trade, as a plasticizer for nitrocellulose lacquer, and in its dehydrated form, as a substitute for Chinawood, oiticica, and other drying oils. It is used to some extent, in soapmaking and in the manufacture of typewriter inks, sticky fly-paper, and lubricants. It is also used as the basis for making capryl alcohol, for perfumery; and sebacic acid, which is used in the manufacture of celluloid plastics.

Castor oil is used in hydraulic fluids. It is the only natural glyceride that is completely soluble in alcohol, which makes it suitable for bandolines and alcoholic hair tonics. It is sold in U.S.P., No. 1, and No. 3 grades.

Cocoa Butter

Cocoa Butter, or, more correctly, *Cacao Butter*, is a pleasant-smelling fat, pressed from the seeds of *Theobroma Cacao*, the chief source of chocolate and cocoa powder. This plant grows profusely in Central and South America. The bean contains about 50% of fat. Some of the beans and press cake are solvent-extracted and used for the production of theobromine for medicinal use, and for cola beverages.

Cocoa butter is used in confections, soap, cosmetics, and as a base for suppositories. Chemical and physical constants are given in Table 3:4.

Coconut Oil

The lauric acid oils, that is, coconut, palm-kernel, babassu, and several others marked with an asterisk in Table 3:4, deserve mention here, because of their importance to the soap industry and their use in the manufacture of lauryl alcohol, on which an important class of surface-active agents is based (see Chapter 5). It may be noted, that these oils as a source for lauryl alcohol and other fatty alcohols, are meeting competition from petroleum-derived fatty alcohols produced by the Alfol process, to be discussed later.

TABLE 3:1

AVERAGE CONTENT OF FATTY ACIDS IN COCONUT OIL

Saturated Acids		*Unsaturated Acids*	
Caproic	0.5%	Palmitoleic	0.2%
Caprylic	8.	Oleic	6.
Capric	7.	Linoleic	2.3
Lauric	48.		
Myristic	17.	Total	8.5%
Palmitic	9.		
Stearic	2.		
Total	91.5%		

Coconut Oil [2] is obtained from the nuts of the coconut palm, which is widely distributed in all tropical countries, especially along the coast lines. The fibrous outer husks of the nuts are used for the manufacture of mats, brushes etc.; the soft part, adhering to the inside of the shell, is the copra, containing approximately 65% oil, excellent for margarine, shortening, soap, and allied purposes. Copra is pressed in expellers. The content of fatty acids in coconut oil is shown in Table 3:1.

Coconut oil is valued by soapmakers, for its high content of lauric acid, which, when it is included in toilet soaps, is the chief suds-producing ingredient. The latter use may also be materially reduced, when more bar soaps are made with synthetic surfactants. The oil is valuable as a base for liquid soap because its potassium soap remains clear in comparatively high concentration (37–40% of anhydrous soap) at a temperature as low at 40°F. This soap may be piped to distant dispensers, without congealing. For use in dispensers, however, it is usually diluted with water to 15% of anhydrous concentration or even lower. It will produce suds quickly, even at low concentration.

There are two grades of coconut oil: Cochin grade, and Ceylon (or Manila) grade.

Cochin Type:

The Cochin type of coconut oil is an alkali-neutralized and bleached oil, nearly water-white, used in high grade soaps, in the cosmetics industry, and in the alkyd industry.

Some constants are:

F.F.A. Max	0.1%
Swift Stability Test	200 hr Min
Color Lovibond 5¼″ Max	10 Y—I R

Other constants are similar to those of the crude oil.

Ceylon Type:

The Ceylon or Manila grade of coconut oil is slightly darker than the Cochin grade.

Some constants of Crude Manila Coconut Oil are:

F.F.A. Max.	6.0%
Iodine No.	7.5–11
Saponification No.	248–254
Sp. Gr.	0.924–0.926
Melting point	72–79°F
Color	20–35 Y.2½–3½ R

Besides the technical grades of coconut oils, just described, there are several grades of edible oils, which will be merely named, because they are of no concern in the present volume.

Edible coconut oil 76° (Congeal Point, 73–76°F)

Hydrogenated coconut oil 92° (Congeal Point 82–86°F; melting point 105–108°F.)

Hydrogenated coconut oil 110° is used similarly to 92°, particularly in the summer season (congeal point, 90–99°F; melting point, 119–125°F.)

Babassu and **Palm-Kernel** oils are used interchangeably with coconut oil in making soap. The other lauric acid oils have not yet attained importance in the United States. They are used, however, in the countries of their origin.

Corn Oil

In the production of vegetable oils, corn oil ranks third in volume, after soybean and cottonseed. Corn oil is obtained from the germ of Indian corn. The cleaned corn is degerminated either by the dry process (used also for making meal, flour, hominy etc.), or by the wet process, which is employed for making starch and glucose.

In the dry process, the corn is cleaned, sprayed with water and steamed to a moisture content of 20%, after which it passes through a degerminating machine, where the hulls and germs are loosened, but the starchy portion is not ground to any extent. The separated germs, dried to 14% of moisture, pass through hominy reels to complete the separation of germs from bran and meal. The final product, containing 18–20% of oil, is sent to the oil mill. About one-half pound of oil is obtained from a bushel of corn.

In the wet process, the cleaned corn is soaked in oblong wooden vats from 30 to 40 hours, in a large quantity of water and 0.2% sulfurous acid. The latter softens the corn and acts as a preservative and bleach. The treated corn then passes through an attrition mill, which shreds the kernels and loosens the germs, and these, by means of agitation, float on the water and pass over the edge of the vat. The germs freed from starch etc., are dried to 5% of moisture. At this stage, the germ will contain 40–50% of oil. The yield is about 1.25 pounds to the bushel of corn, degerminated by this process.

The oil is expressed by the Anderson Expeller, or it may be continuously extracted with solvents. It is refined, bleached, and winterized, as is cottonseed oil. It is sold for food in this form, or it may be hydrogenated (see Chapter 6). It is also used in oleomargarine, because it is high in polyunsaturated (see Chapter 4).

Cottonseed Oil

Second in importance to soybean oil, is cottonseed oil. The cotton plant is cultivated in U. S., India, China, Russia, Brazil, Egypt, Mexico, Uganda, Peru, Turkey, and other countries; in fact, more than seventy-five countries cultivate it.

Crude cottonseed oil, as obtained from the seed, has a strong, characteristic odor and dark, reddish-brown color from the resinous and other highly colored material in the seed. American crude oils vary from 0.3 to 3% of free fatty acids.

This oil contains more saturated fatty acids than most oils of an equivalent iodine number and therefore the titer is high, (33 to 36°C). Stored at 50 to 60°F, it may become partially solidified. It is classed as a semi-drying oil.

The approximate percentage composition of the fatty acids is:

Saturated	25
Oleic	25
Linoleic	50

There are two grades of cottonseed oil: Prime Summer Yellow and Winter Oil. Most of the stearin has been removed from the latter, so that when chilled, it does not deposit stearin. The oil is used chiefly for edible purposes. The soap-stock may be either converted to cheap soap powder; or acidulated and distilled, then used for various industrial purposes.

Cottonseed meal, after it has been freed from gossypol, is used for animal fodder.

Linseed Oil

Flaxseed, from which linseed oil is pressed, is grown almost all over the world, with Argentina, United States, Russia, India, and Canada as the largest producers. China, Morocco, and Uruguay also produce flaxseed.

In Russia, considerable quantities of flaxseed are cold-pressed to produce edible oil. In the manufacture of industrial oil, the crushed seeds are *cooked*, that is, heated at 200°F to reduce moisture and to bring the seeds to proper consistency for maximum yield. The seeds are then pressed in hydraulic presses, yielding about 34% oil.

The pressed oil contains phosphatides and mucilaginous impurities.

When heated to 350°F it becomes cloudy; that is, it "breaks." The cloud is filtered out and, during continued settling, a clear oil is obtained, which is known as *raw linseed oil.* It may be further refined to remove color and the remaining mucilaginous matter. The oil may be bleached by heating to 212°F with 2–5% of fuller's earth, and filter-pressed.

So-called "boiled linseed oil" dries more readily than the raw oil does. In producing boiled linseed oil, driers consisting of the cobalt, lead, and manganese salts of linoleic, naphthenic, and rosin acids are added while the oil is being heated to 350°F in open kettles.

Pale or bleached "boiled" oil is produced by heating the oil and drier to a temperature not exceeding 250°F. A gentle current of air is blown through the oil. It is then quickly cooled to 150°F and more drier is added, followed by a short but vigorous blowing. The oil is allowed to settle and is then filtered.

Blown linseed oil of almost any desired viscosity can be produced by blowing with air and heating to even as high as 550°F; (see Chapter 7). Aluminum kettles are employed and the heating may be effected with steam, Dowtherm, or electricity. The flammable fumes are led away through stacks. "Stand" oil is oil that was heat-bodied without driers.

Blown linseed oil is used with other oils to promote the wetting of pigment in paints and enamels. It is also serviceable in the manufacture of lithographic varnish, oilcloth, and patent leather.

When linseed oil is heated at 300°C in the absence of air, its iodine number begins to fall rapidly. Soon the specific gravity increases, and the oil becomes viscous. During the later stages of heating, the increase in viscosity becomes more pronounced, and the iodine number continues to fall, but at a slower pace. This is explained by the fact that some of the ethylenic linkages become saturated, not by the addition of hydrogen or oxygen, but by some kind of polymerization.

Heat-polymerized linseed oils are used in paint and varnish because of their levelling properties and gloss.

Olive Oil

Olive oil, also called *sweet oil,* is a yellow or greenish-yellow liquid oil from the fruit of a tree, *Olea Europea.* Its specific gravity is 0.910–0.918, and it is soluble in most common organic solvents. (For other properties, see Table 3:2.)

Spain, Italy and Greece produce the finest olive oil, but many other countries bordering on the Mediterranean Sea are also producers. In recent years, a branch of the industry has developed in California and Arizona in the United States.

The manner of producing olive oil is somewhat different from that for other oils.

It has been found that the content of oil is greatest before the olive ripens. The green fruit is gathered, cleaned, and then crushed in stone mills. The pulp is pressed by hydraulic pressure, the virgin oil is allowed to settle, and the clear oil is decanted for packing. This first-grade olive oil is used as food and in medicaments.

The press cake is alternately ground and pressed several times to recover the lower-grade oil, which is known as *industrial oil* and is used for sundry purposes, such as soapmaking, oiling silk and wool, etc. It may also be sulfonated.

The residual press cake is then subjected to extraction with carbon disulfide and is marketed as *sulfur olive oil*, or *olive oil foots*. This is a viscid, foul-smelling green or pale-green oil. Its content of free fatty acids may run as high as 70%. It contains chlorophyll and cellular matter from the pulp. It is used chiefly in the manufacture of textile soap. When converted to toilet soap, it is known as *green castile* soap (see Chapter 7.)

Palm Oil

There are several grades of palm oil. When the fresh palm nuts are pressed, the yield is an orange-colored oil with low content of free fatty acids. In temperate climate, the oil is viscid to solid. This grade is used for edible products. As most palm oil, however, is pressed by natives in the East Indies or Africa using slow primitive methods, it comes to market in more or less decomposed form, containing 40–60% of free fatty acids.

Lagos and Niger palm oils are used in soapmaking. It has been used as a flux in tin plating, but is almost entirely displaced by other fatty acids, at present.

Safflower Oil *

Safflower oil is obtained by pressing or solvent extraction from the seeds of *Carthamus Tinctorius* (*American Saffron*, *False Saffron*). It is grown in the Middle and Far East, but in greatest amount in the United States. In general, it is somewhat similar to linseed oil.

* Suppliers: 62, 108, 168.

It has been reported that farmers planted 500,000 acres of safflower in 1962—20% more than in the previous year. The production of oil for 1963 was 116,000,000 pounds, of which 76,900,000 was refined. Production of refined oil for 1964 was 75,400,000 pounds.

It is estimated that 80% of the crop comes from California and Arizona, 13% from Nebraska, Colorado, and western Kansas, and 7% from Montana and Wyoming.

The great impetus in the growth and production of safflower is due to medical announcements about the relationship between the intake of ordinary—that is, "saturated" **—animal fats and the amount of cholesterol in the blood. As a result of long-continued scientific and medical research, it is believed that if "unsaturated" ** ("polyunsaturated" **) vegetable oils are substituted for animal fats and saturated vegetable fats, there will be less cholesterol in the blood, and, in consequence, less danger of arterioscelerosis, heart attacks, and strokes.

Top quality seed now yields 38% of oil, as compared with the 33% obtainable a few years ago. New seeds under cultivation have been reported to yield 40% of oil. Its glyceride content is: Saturated, 8.6%; oleic, 16.7%; linoleic, 71.3%; linolenic, 3.4%.

Safflower oil has 76% of "polyunsaturates," which is considerably higher than the figure for corn oil—55%.[00] Constants are:

Specific gravity	0.923–0.928
Refractive index (40°C)	1.4679–1.4693
Solidifying pt.	−13– −20°C
Saponification No.	187–194
Iodine No.	130–150
Unsaponifiable	0.5–1.5%
Titer	11–17°C
SCN value	82.2

This oil can be used industrially in paints because of its high iodine number; it is also used in varnishes and linoleum. It is being promoted however, for sale to cholesterol-minded householders as an oil with the highest "polyunsaturates"—therefore, valuable in foods and special diets.

Although a great amount of interest and research has been devoted to the polyunsaturated portion of several oils, there is still considerable uncertainty about the exact value of polyunsaturated fats in the diet.

** For explanation of these terms, see Chapter 4.

BHA (butylated hydroxyanisole), BHT (butylated hydroxytoluene), propyl gallate, and citric acid—propylene glycol are used as antioxidants ("freshness preserved with") according to the label on Saff-O-Life (Betty Crocker; General Mills) a product for salads, baking, and frying. It is also used in margarine, Golden-Glo (Lever Bros.). Moreover, a well known dog food has been advertised as containing polyunsaturated oils.

Soybean Oil

In point of volume of production, soybean oil ranks first among the vegetable oils produced in the United States. The bean yields 15–20% of oil, which may be produced by solvent-extraction or by pressure. The expressed oil "breaks," when heated but the material that breaks out can be completely removed by alkali-refining.

Solvent-extracted oil, usually gives no "break." The beans are cleaned, dried and flaked, then treated with the solvent. The solvent is distilled off and that recovered from steaming the meal is returned to the solvent storage tanks. The meal is dried to the desired moisture

TABLE 3:2

COMPARABLE TALL OILS AND TALL OIL FATTY ACIDS

Product	*% Rosin*
Acintol D	27–31
Acintol FA 3	0.6
(Arizona)	
Unitol K	25–28
Unitol R	39–42
Unitol S	40–43
(Union Bag-Camp)	
Indus oil M28	25–28
F.A. Specialty 57–118 and Indus oil R90	Negative Liebermann-Storch reaction for rosin
(W. Va. P.P.)	
Sylvatol 40 DD	28
Sylfat 500, 496, 96	1
(Glidden)	
Pamak F.A. 25	27
Pamak F.A. 1	1.3
(Hercules)	
Acosix or Acosix 700	25
Aconew 500	1
(Newport)	
Facoil	
(Natl. Sou.)	
Emtall 662	49
(Emery)	

content. It contains about 1% of oil. It also contains 41–50% of protein. It is used for human food, as feed for live stock, and in the manufacture of adhesives and plastic compounds.

Tall Oil

Before World War II, so-called *tall oil* did not receive much attention in the United States. Because of shortages, however, and the rising prices of other oils about that time, much research was devoted to this commodity. It has been very extensively exploited as a substitute for other oils and fatty acids, so that by the end of the war, the supply was unequal to the demand.

Tall oil, although of vegetable origin, is not produced in the manner of other oils, nor is it a true oil. It is obtained by the acidulation of "black-liquor soap," which results from cooking pine-wood chips, in the manufacture of sulfite paper. Crude tall oil is a brownish-black, viscid oil, which consists of about 45% of rosin and 50% of linolenic, oleic, and linoleic acids, with 5% or more of sterols and impurities. It has been separated by fractional distillation in vacuum, to yield a fatty acid fraction of about 99% and 1% of rosin; and, vice versa, a product containing 95% rosin with a low fraction of the fatty acids (Table 3:2.)

Fractionation of this sort is relatively expensive and even with the added cost, distilled tall oil is priced considerably lower than other fatty acids.

A sample analysis of crude tall oil follows:

Moisture and volatile matter	0.90%
Ash	0.04
Unsaponifiable fat	6.50
Free fatty acids (as oleic)	85.95
Acid number	172
Saponification number	174
Titer	about 10°C
Fatty acids	49.52
Rosin acids	42.00
Viscosity (Saybolt Universal, 210°F)	116
Iodine number (Wijs)	159
Flash point (Cleveland open cup)	350°F

Tall oil can be used in the manufacture of soaps for treating textiles, paper, cleansers, cutting compounds, metal-drawing and metal-forming compounds, agricultural adhesives and emulsifiers. The well-known surfactants, the Renex 20 series (Atlas), are nonionic esters formed by reacting ethylene oxide with refined commercial mixtures of unsaturated fatty acids and heterocyclic resin acids (refined tall oil).

In its sulfonated form, tall oil has been found useful in cold water or casein paints. Esterified with pentaerithritol, it is used in varnishes.

Miscellaneous Vegetable Products

Chinese Vegetable tallow: This is produced by melting the fat from the fruit of the stillingia tree, which grows in China, southern Japan, islands of the Pacific, California, South Carolina, Florida, Georgia, and Texas. Very little has been imported in recent years. It is used for making candles.

Crambe Seed Oil:

Crambe (or *Crambe Abyssinica*) is an annual herb which grows wild around the Mediterranean region. The U.S. Department of Agriculture, in cooperation with Pacific Vegetable Oil Corporation, has carried out full-scale run at Sidney, Nebraska, on 36 tons of crambe seed.[3]

Field tests have been carried out for the past three years, to develop crambe seed as a domestic industrial oil crop. These tests, carried out in Texas, Nebraska, Iowa, and Montana, indicate that the plant grows well in several areas of the United States.

Work at Peoria, on the meal, is aimed to develop a practical method for producing a bland, palatable product for live-stock feed.

Crambe seed is similar to rapeseed, but it contains more erucic acid. The seed is cleaned, cracked and the hulls are removed. The meats cut in 1–10 inch diameter are passed through a conditioning oven and thence through an expeller, where 75% of the oil is pressed out. The press-cake is extracted with hexane to remove the remaining oil, after which the solvent is flashed off with steam.

The hull-free seed contains 27% protein and 46% oil. Its composition is 59% erucic, 18% oleic, 10% linoleic, 5% linolenic and 8% miscellaneous fatty acids.

Erucic acid can be cleaved into new dibasic acids for the manufacture of resins, plastics, fibers and coatings.

Ucuhuba Butter:

Before World War II, quantities of ucuhuba butter were imported from Brazil. It is valuable as a source of myristic acid, of which it contains 63 per cent. Soap made from this fat produces a very good lather.

Many South American and Asiatic oils and fats have been included in Table 3:4 in view of their possible future exploitation. Most of them are used in the countries of their origin.

Rosin

As far as the soap industry is concerned, rosin has lost its importance. Because of the rise in price of rosin, which now costs considerably more than tallow and other saponifiable fats, its use in laundry soap has been vastly curtailed. Several brands of rosin soap are still being produced, as described here, but the modern housewife now depends more on a washing machine, with detergents, than on the washboard where her mother and grandmother used rosin soap in bar form. By using a small amount of crude tall oil with hard fats, the soapmaker can produce a rosin soap which has the appearance of the old time bar of soap. And so, rosin is used for other purposes than in soap.

Gum rosin is obtained from three species of pine trees indigenous to the Carolinas, Georgia, and Florida. An incision is made in the growing tree and the sap, which consists of turpentine and rosin, flows into a cup suspended under the incision. The sap is collected every two weeks, and taken to the stills. The turpentine is separated by steam distillation and the rosin remains in the still. This is strained, sometimes bleached, and run into barrels where it hardens and is ready for shipment.

Gum rosin varies in color and is graded down from WW (water-white), to A (black). The lowest grade acceptable to soapmakers is K, of a transparent amber-reddish color.

The so-called *wood rosin*, although it is also obtained from the pine tree, is recovered in a different manner. The rosin is extracted from stumps and dead branches. These are chipped up and charged into retorts or stills employing superheated steam. Pine oil, turpentine, and rosin are recovered from the distillate. Many derivatives are obtained from the terpene portion of turpentine.

Rosin has approximately the same saponification number as tallow. Its iodine number is 200–228. Its chief ingredient is an acid, having the empirical formula $C_{20}H_{30}O_2$, which may be considered as abietic acid or one of its isomers. Rosin contains about 6% of unsaponifiable matter.

Rosin is seldom used alone in the production of soap, because its sodium salt is very soft and sticky, although one firm has succeeded in producing a resinate in powdered form. A powdered tall oil soap is also being sold. Rosin is usually combined with tallow, house grease, or other hard fat and saponified with caustic soda to produce the common yellow bar of scrub soap. This type of soap has good detergency, especially with added builders, such as silicate of soda and soda ash. It dissolves readily and produces abundant suds. This type of soap, with naphtha added is still being produced.

Besides its use in soap, rosin has a very definite place in the manufacture of varnishes where it is applied in the hardened form, by combination with lime. Reacted with glycerol it is known as *ester gum*, which is used in paints and varnishes. The methyl ester of rosin is a viscous liquid which is used as a plasticizer. Zinc resinate is used with linseed oil in traffic paints, and in resin-oil emulsion paints.

BY-PRODUCTS AND DERIVATIVES

Soapstock or Foots

The residue remaining in the refining tank, from cottonseed, soya, and corn oils, after the oil is drawn off, which is commonly called *soapstock* or *foots*, is used in the manufacture of soap powder. It may be used "*as is*," or it can be acidified with sulfuric acid and sold as *acidulated soapstock* or *acidulated fatty acids*

The foots consists of soap, occluded neutral oil, gummy cellular matter, and dirt and water from the lye that was used in refining. When sold "*as is*," the foots contains about 50% of fats. By acidulating and separating the fatty acids from most of the extraneous substances, the fatty acid content can be increased to 95–98%, and therefore about half of the space and weight is conserved in shipping. Much of this type of fatty acid is steam-distilled to improve color and quality. Some types are pressed to obtain red oil and stearic acid.

Some refiners sell boiled-down soap. This is obtained from the foots by the following process. The foots, which is only partially saponified as a result of the refining process, is completely saponified and given one or two washings with salt to remove gums and dirt. It is then boiled with an excess of salt. This treatment concentrates the soap and at the same time, hardens it so that it can be shipped in slack barrels or even in paper bags. It is sold on the basis of 65% total fat content. It can be used in scouring wool, in the production of soap powder, and where a cheap grade of soap is satisfactory.

Another way of disposing of the foots is to subject it to the Twitchell process (see Chapter 6), during which the neutral fat is split into fatty acids and glycerol. The fatty acids are then distilled, leaving a residue of pitch which may be sold for roofing, insulation, and other purposes. The by-product glycerol ("sweet-water") is neutralized and evaporated to crude glycerol and then distilled.

Oil Seed Meal

Besides the oil produced by crushing, pressing or extraction, the most important product of usable seeds is the meal. In nearly every case the meal, which is rich in protein, may be used for animal feed; or, as with soybean, the flour is used for human and animal food, and for the production of adhesives and plastics.

Cottonseed meal contains a toxic substance called *gossypol.* By cooking the seeds with at least 14.5% of moisture at 115°C for 90 minutes, the free gossypol is destroyed and the product is then suitable for animal feed.

Castor seed press cake is used mostly for fertilizer. If its toxic albumins, ricin, and ricinine are extracted several times with hot water (according to a German process), it may then be used as fodder. Ricin is used as a reagent for pepsin and trypsin.

Almond meal is used in cosmetics, perfumery, and confectionery.

Lecithin

Lecithin * is a fatty acid ester of glycerophosphoric acid and choline. The fatty acids contained in lecithin are stearic, palmitic, oleic, linoleic, and linolenic acids. Although lecithin is found in most animal and vegetable oils and in the brain, nerves, blood plasma, bile, egg yolk, and fish eggs, it is recovered commercially only from egg yolk, soybean oil, and to a small extent, from corn oil.

Soybean oil contains about two per cent of lecithin, but the actual amount recovered is about one-half per cent. The phosphatides are separated from the oil by the introduction of steam, which hydrates and consequently precipitates them out of the oil. The lecithin is then recovered by centrifuging.

By another method, lecithin is produced from soapstock, the by-product of alkali refining of soybean oil or corn oil. The soapstock is treated with petroleum ether in which the soap is insoluble, but which acts as a solvent for the phosphatides. Lecithin is then recovered from the solution by distilling off the solvent.

* Suppliers: 6, 13, 168.

Soy lecithin is sold as a greasy, yellowish-brown substance, containing about 30% of soybean oil. The confectioner's grade contains 30% of cocoa butter.

Lecithin is soluble in oil, alcohol, benzene, ether, and chloroform. It is practically insoluble in acetone, ethyl acetate, and similar esters. It is the only oil-soluble edible colloid obtainable in commercial quantities.

In vegetable butters, shortenings, and margarines, lecithin helps to simulate the natural characteristics of dairy butter. It increases homogeneity and uniformity of breakdown on melting, and prevents nonspattering in frying. Candies and baked goods are improved by the addition of lecithin. It is used in chocolate coatings to prevent blooming.

Lecithin is also used to depress surface tension, to control viscosity, and as a dispersing, emulsifying, and foam-producing agent. It increases the cleansing and softening effects of cosmetics and soaps. It serves as an antioxidant in finishes for leather and textiles and in pharmaceutical preparations.

About 35 million pounds of lecithin are produced annually in the United States, of which 15 million pounds are exported. Its uses in food and for industrial purposes are listed in Table 3:3.

TABLE 3:3

COMMERCIAL DISTRIBUTION OF LECITHIN

Industry	*Million lb*
Margarine	5.5
Chocolate	3.5
Baked Goods	1.5
Animal Feeds	0.5
Paint Products	3.0
Printing Inks	0.5
Petroleum Industry	1.5
Soaps, Cosmetics	0.5
Miscellaneous	3.5
Total	20.0

Phytosterols

Plant sterols may be obtained from the oils in which they occur by several methods, one of which is molecular distillation. The sterols are also extracted from the unsaponifiable matter, after saponification of the oil, followed by fractional crystallization.

In still another process, sterols are recovered by mixing soybean oil with 8% of adsorbent such as aluminum silicate, (based on the weight of oil). The mixture is recirculated through a filter press. The cake is then removed from the press and extracted with acetone. The solvent is evaporated, producing a concentrate which contains 0.6% of sterols, which are further concentrated by extraction with methanol. A portion of the sterols crystallizes out. The balance is evaporated to remove the alcohol. The residue is saponified and the remaining sterols are extracted.

Stigmasterol, discovered by Hesse in 1878, was derived from soy and clabar beans, and has been transformed in the laboratory into the hormones progesterone and desoxycorticosterone.

Pure stigmasterol consists of colorless needles, melting at 170°C insoluble in water, but soluble in most organic solvents.

The chemical and physical constants of most of the world's supply of vegetable fats and oils are summarized in Table 3:4.

Besides classification as to origin, vegetable oils are also subclassified according to their behavior when exposed to air. From this viewpoint, they are called drying, semidrying, and nondrying oils.*

Drying oils, when spread in thin layers, soon form films that become hard. For this reason, they are used in paint and varnish where they act as vehicles for pigments.

In this class are China-wood (tung), hempseed, linseed, oiticica, perilla, sunflower-seed and poppy-seed oils. More recently, dehydrated castor oil (see Chapter 6) and some "tailor-made," solvent-fractionated fish oils, which have good drying properties, have been included in this category.

Semi-drying Oils are used mainly in edible products. Corn, cottonseed, rapeseed, sesame, safflower and soy-bean oils are important members of this class. Much soybean oil is used also in combination with linseed oil, for paint and varnish. Corn, cottonseed and soybean oils are treated with sulfur chloride for making factice, which is a rubber substitute. Rapeseed oil is used in marine lubrication.

Non-drying Oils: Almond, castor, coconut, olive, palm oil, palm kernel and peanut oils are best examples of vegetable nondrying oils. They are used as edible oils (except castor), in cosmetics and in various other fields.

* Suppliers: 13, 19, 27, 136, 141.

TABLE 3:4

CHEMICAL AND PHYSICAL CONSTANTS OF OILS AND FATS

Name	*Source*	*Type*	*Saponification No.*	*Iodine No.*	*Titer °C*	*Melting Point, °C*	*Percent Yield*
Almond	World	n.d.	183–207	93–103	12		40–61
Andiroba	W. Indies, Brazil	n.d.	195–6	58–78	35–37		13
Aoura Kernel	Central, South America	n.d.	242–9	10–11	26	29–30	46
Mexican Poppy or Argemone	Mexico	s.d.	193	124–8	23		36
Babassu *	Brazil	n.d.	247–50	14–16	22		63
Bacaba	Guiana, Tropics	n.d.	192	78			
Bacury Kernel	South America	n.d.	199.5	78		51	66
Baratinha	South America	n.d.	181	78			60
Butter Fat			227	26–50	33–38	Polenske No. 2–3 Reichert-Meissl No. 24–32	
Butternut	Brazil	s.d.	199	49.5	23–29		
Cacao Butter	South America, W. Indies	n.d.	191	41	48.4		
Cashew	Brazil, W. Indies	n.d.	195	84			
Castor	U.S., Haiti, S. America	n.d.	177–87	82–90			45
Chicken Fat			193–205	66–71	32–34	33–40	
Chinese Vegetable Tallow	China, India	n.d.	179–206	28	51	41–60	60
China-Wood *See Tung*							
Coconut *	Tropics		251–64	8–10	23		63
Cod Liver	Norway		171–89	137–67	21		
Coffee	Brazil, British Honduras	n.d.	165–77	85–90		38–40	9
Cohune *	Guatemala, Mexico	n.d.	251	12–13	21	27–30	40
Copaiba	South America	d.	77.8	174			40
Corn (Maize)	U.S.	s.d.	187–93	111–28	20		
Cottonseed	U.S., Mexico, South America	s.d.	192	103–15	32–38		15.4–28
Coyol	Central America, Mexico	n.d.	246	25		25	48
Dende	Brazil	n.d.	199	78–83		22–30	
Gergelim	South America	s.d.	187	104			
Haddock Liver			189	154			
Hog Grease			195	60–68	34-38		
Illipe Butter (Mowrah)	India		188–207	53–70		24–25	50–60
Jaboty	South America	n.d.	228	23		45	42–53
Jupaty			194	77			
Kapok	East & West Indies, Africa	n.d.	189–96	74–117	27–38		18–25
Lard			195	56	34–42		
Licury	South America	n.d.	233	23		45	
Linseed	U.S., USSR, South America	d.	189–96	179–204	19–21		34
Mahuba	Brazil		252	18	40–44		50
Mamorana		n.d.	176–86	81–90			
Marcita	New Guinea	d.	203–214			37–50	12
Maripa	W. Indies	n.d.	256–70	4–17		18–27	
Menhaden			189–93	148–85			
Mocaya * (Gru-gru)	Trinidad, Paraguay		240	16–24	20–23		60
Muriti Fat	West Indies, Brazil		246	25	17	25	50
Mustard	Europe	s.d.	174–80	95–106	6–8		33

* Belongs to the coconut oil group.

TABLE 3:4 (*Continued*)

Name	*Source*	*Type*	*Saponification No.*	*Iodine No.*	*Titer °C*	*Melting Point, °C*	*Percent Yield*
Nutmeg Butter	Brazil, Guiana, W. Indies	n.d.	174	57		25–48	38–40 (12% Essential)
Neem (Margosa)	India, Burma		186–204	70–73	42		
Ocuba Fat	Brazil					39–40	
Oiticica	Brazil	d.	188–94	145–79			55–62
Olive	Spain, Italy, South America	n.d.	185–200	74–94	18–25		25–30
Palm	Africa, E. Indies		200–205	49–59	45		
Palm Kernel *			248	13	23		50
Patua (Batua)	Brazil	n.d.	196	75	17		24
						Resembles Olive Oil	
Peach Kernel		n.d.	191–93	92–100			
Peanut	U.S., Argentina, Mexico	n.d.	186–94	88–98	27–30		40–50
Perilla	China, Japan, Manchuria	d.	190	205			38
Piquia	Costa Rica	n.d.	200	26			
Pracachy	Brazil		170–7	69			51
Rapeseed	Manchuria, Argentina	s.d.	168–80	98–106	12		35
Rosin	U.S.		175–90	220–228	20		
Rubberseed	Amazon Basin	d.	185–206	117–33			
Safflower	Middle East, N. Africa, S. Eur.	d.	190	150	16–17		24–37
Sardine	U.S.		191	160–190	28		
Seal			189–96	127–141	16		
Sesame (Benne)	Mexico, Cuba, Argentina	s.d.	188–193	103–115	21–24		47
Shea Butter	Africa		178–89	56–65	52–53		
Soybean	U.S., Asia	s.d.	189–193	122–35	20		15–20
Sperm			123–147	80–84			
Sunflower	USSR, Hungary, Argentina	s.d.	189–94	120–36	17–20		45–50
Tall	U.S., Scandinavia		175	170–185	10		
Tallow			190–96	55	40–46		
Teaseed	Guatemala, Asia	n.d.	197	85	20		30–45
Tonka Butter *	Guiana, W. Indies	n.d.	257			28	
Tucuma (Aoura)	Guiana	n.d.	220	46	27–30		44
Tucuma Kernel *	South America	n.d.	220	11.6		30.5	52
Tung (Chinawood)	China, Brazil, Argentina	d.	189–95	160–170	37–38		
Ucuhuba **	Brazil	n.d.	219–21	9–14	45		
Uricury	W. Indies, Brazil	n.d.	242	12.8	25–36		
Whale		n.d.	197	90–146	23		

Abbreviations: d. Drying; s.d. Semidrying; n.d. Nondrying.

** Contains 12.6% lauric acid and 63.2% myristic acid.

For identification of all fats and oils mentioned in this book, according to this classification, see Table 3:4.

REFERENCES

1. German Patent, 698 200.
2. Leaflet published for the trade; Welch, Holme & Clark Co., New York, 1964.
3. *Chem. Eng. News*, March 2, 1964; August 2, 1965; January 1, 1966.

chapter 4

FATTY ACIDS—FATTY ALCOHOLS—GLYCEROL

The fatty acids are found in nature, combined with glycerol, in the form of esters, also known as *triglycerides.* Most oils and fats consist of glycerides of the higher fatty acids, such as lauric, myristic, palmitic, oleic, stearic, linoleic, etc. The lower members of the series such as butyric, caproic, caprylic (containing four to ten carbon atoms) are found in small quantities in such oils as coconut, palm kernel, and babassu. Butyric glyceride is found in milk fat. (See Table 4:1.)

FATTY ACIDS

On a commercial scale, the fatty acids are derived from the glycerides by hydrolysis. For every three molecules of fatty acids, one molecule of glycerol is split off. The fatty acids are then refined or distilled to a purity which is almost equal to that of the chemically pure products.

With advances in chemical technology, the individual fatty acids were required in pure form. Table 4:2 shows how closely two commercial fatty acids compare with pure stearic and pure lauric acids. The other individual fatty acids are equally good in comparison with their pure counterparts. Practically all the producers of fatty acids now offer such pure products.

The fatty acids, much like other organic acids can undergo such reactions as oxidation, hydrogenation, dehydration, polymerization, condensation, and conversion into salts.

Derivatives of fatty acids are available in tank-car quantities. Fatty alcohols, mono- and polyesters, their sulfates, sulfonates, acid chlorides, amines, amides and nitriles have found use in the manufacture of surfactants, emulsifiers, plastics, plasticizers, flotation agents, finishes for textiles and leather, cosmetics, and pharmaceuticals.

TABLE 4:1

NATURALLY OCCURRING SATURATED FATTY ACIDS

($C_nH_{2n}O_2$)

Common Name	*No. of Carbons*	*Chemical Formula*	*Mol. Wt.*	*M.P., °C Pure Acid*	*Occurs in*
Butyric	4	C_3H_7COOH	88.1	−8	Milk Fat
Caproic	6	$C_5H_{11}COOH$	116.16	−2	,, ,,
Caprylic	8	$C_7H_{15}COOH$	144.21	16.0	,, ,, Coconut Palm Kernel
Capric	10	$C_9H_{19}COOH$	172.26	31.3	Coconut, Palm Kernel
Lauric	12	$C_{11}H_{23}COOH$	200.31	43.6	Coconut, Palm Kernel Babassu
Myristic	14	$C_{13}H_{27}COOH$	228.37	54.0	Animal Fats, Nutmeg, Ucuhuba
Palmitic	16	$C_{15}H_{31}COOH$	256.42	63.0	Animal, Vegetable Fats
Stearic	18	$C_{17}H_{35}COOH$	284.47	69.6 (Coml 53–55)	Animal, Vegetable Fats
Arachidic	20	$C_{19}H_{39}COOH$	312.52	75.3	Peanut
Behenic	22	$C_{21}H_{43}COOH$	340.57	80.0	Peanut, Rape, Mustard
Lignoceric	24	$C_{23}H_{47}COOH$	368.63	84.0	Peanut, Rape, Mustard
Cerotic	26	$C_{25}H_{51}COOH$	396.67	85–87	Beeswax, Carnauba

TABLE 4:2

COMPARISON OF FRACTIONATED AND PURE FATTY ACIDS

	*Neo-Fat 18**	*Stearic Acid*	*Neo-Fat 12–43**	*Lauric Acid*
Mean Molecular Wt.	284.3	284.28	200.5	200.19
Melting Point, °C	67.0	69.9	43.5	43.6
Iodine No. (Wijs)	0.7	None	0.2	None
Neutralization Value	197.0	197.0	279.8	280.0
Unsaponifiable Matter	0.3	–	0.1	–
Color	White	White	White	White
Odor	None	None	None	None

Compositions:

Capric	– %	0.3%
Lauric	–	99.0
Myristic	–	0.7
Palmitic	5.6	–
Margaric	2.5	–
Stearic	91.2	–
Oleic	0.7	–

* Products of Armour & Co.

Saturation in Fatty Acids

The type of glyceride occurring in natural fats and oils depends on the kinds of fatty acids linked with the glycerol. The accepted formula for fatty acids consists of a chain of carbon atoms, each of which carries two hydrogen atoms, except the first, which has three (methyl group, CH_3), and which is connected with the glycerol.

When the fatty acid portion of a formula contains all the hydrogen it can carry it is called *saturated.* The formula of a typical saturated fatty acid is that of lauric acid (in coconut oil):

$$\begin{matrix} & H & & H & & H & & H & & H & & H & & H & & H & & H & & H & & H & & \\ H & - & C & - & C & - & C & - & C & - & C & - & C & - & C & - & C & - & C & - & C & - & C & -COOH, \text{ or } CH_3(CH_2)_{10}-COOH \\ & H & & H & & H & & H & & H & & H & & H & & H & & H & & H & & H & & \end{matrix}$$

A fatty acid that does not contain its full quota of hydrogen atoms is called *unsaturated.* This is shown in the formula by a double bond between two of the carbon atoms. In brief form, the formula for oleic acid is

$$\begin{matrix} H & & H & & H & & \\ HC & -(CH_2)_7- & C & = & C & -(CH_2)_7- & COOH \\ H & & & & & & \end{matrix}$$

Such an acid, containing only one double bond and lacking only two hydrogen atoms, is called *mono*-unsaturated (or just *unsaturated*); see Table 4:3.

TABLE 4:3

UNSATURATED FATTY ACIDS

Common Name	*Carbon Atoms*	*Chemical Formula*	*Melting Point °C*	*Occurs in*
WITH ONE DOUBLE BOND, $C_nH_{2n-2}O_2$:				
Myristoleic	14	$CH_3(CH_2)_3CH{:}CH(CH_2)_7COOH$		Milk fat, sperm whale
Palmitoleic	16	$CH_3(CH_2)_5CH{:}CH(CH_2)_7COOH$	−1.5	Milk fat, marine animals, seed fats
Oleic	18	$CH_3(CH_2)_7CH{:}CH(CH_2)_7COOH$	13.0	Vegetable, animal oils and fats
WITH TWO DOUBLE BONDS, $C_nH_{2n-4}O_2$:				
Linoleic	18	$CH_3(CH_2)_4CH{:}CHCH_2CH{:}CH(CH_2)_7COOH$	−5.0	Most seed oils
WITH THREE DOUBLE BONDS, $C_nH_{2n-6}O_2$:				
Linolenic	18	$CH_3CH_2CH{:}CHCH_2CH{:}CHCH_2CH{:}CH(CH_2)_7COOH$	−11	Linseed, soy, perilla
Elaeo-stearic	18	$CH_3(CH_2)_3CH{:}CHCH{:}CHCH{:}CH(CH_2)_7COOH$	49	Tung

Those acids that have more than one double bond—each one of which means two more hydrogen atoms missing—are called *poly-unsaturated.* In such acids, the degree of unsaturation may involve 2, 3, 4, or more linkages in the chain, with, correspondingly, 4, 6, 8 or more hydrogens missing.

Linoleic acid, $C_{18}H_{31}COOH$ is "polyunsaturated." There are some that contain three, four, five, and even six double bonds. Arachidonic acid ($C_{19}H_{31}COOH$), which contains four double bonds, occurs in brain, liver, blood, and in the depot fats of cattle and hogs. Culpanodonic acid, $C_{21}H_{33}COOH$, found in all marine animal oils, has five double bonds. Nisinic acid, $C_{23}H_{35}COOH$, found in Japanese sardine oil, has six double bonds.

Nutritionally, the position of the unsaturation in the formula may determine the point of breakup of the chain in metabolism and how well the body can handle the remaining fragments.

The typical characteristics of several natural fats and oils, and the fatty acids of which they are composed, are shown in Table 4:4.

Synthetic Fatty Acids

In 1882, Schaal was granted a patent for the preparation of fatty acids, by the oxidation of hydrocarbons. Although this process was developed in Germany, some similar work has been done also in the United States.

Noller and Bannerat,[1] in 1934, synthesized oleic acid, and in 1936 Kuhn *et al* produced fatty acids from crotonaldehyde.

In 1942, a patent granted to J. J. Owen, covered a process for the preparation of light-colored fatty acids by subjecting petroleum wax to limited catalytic oxidation in the liquid phase, at temperatures between 100 and 150°C. The product of this reaction is saponified with alkali at 150°C under pressure. The unsaponified matter is separated from the soap by extraction. A small amount of sodium persulfate is added to the crude soap, which is heated at 90–95°C for one-half to two hours, to bleach it, and then it is quickly acidified with dilute sulfuric acid, to separate the fatty acids.

According to Wittka,[2] the problem of synthesizing fatty acids has been solved, but he suggests that the following three processes should be considered for commercial production:

1. Potash fusion of chlorinated hydrocarbons;
2. High pressure hydrogenation of carbon monoxide;
3. Saponification of aliphatic nitro compounds.

TABLE 4:4

TYPICAL CHARACTERISTICS AND CONTENT OF FATTY ACIDS IN NATURAL FATS AND FATTY OILS *

			Babassu	Castor	Coconut	Corn	Cottonseed	Linseed	Olive	Palm	Palm Kernel	Peanut	Sesame	Soya	Tallow	Whale	Herring	Rapeseed
Specific Gravity at 15°C			0.87 (100°F)	0.96	0.93	0.92	0.92	0.93	0.92	0.92	0.87 (100°F)	0.92	0.92	0.92	0.94	0.92	0.92	
Saponification Number			246–250	177–187	251–263	186–193	191–196	189–196	189–195	196–205	244–255	185–192	188–193	191–194	193–200	188–194	180–190	171–183
Iodine Value			12–20	83–86	8–10	120–130	105–115	170–204	79–86	48–58	16–23	83–95	103–115	125–140	35–47	110–140	135–145	95–105
Titer (°C)			21–25	3	20–23	15–19	32–38	19–21	17–21	42–45	20–25	28–30	20–24	23–24	41–45	22–24	22–26	12–15
Fatty Acid	*No. C Atoms*	*Doub. Bonds*	*Fatty Acid Content—Percent by Weight*															
Caproic	6	0	0.3		1.0						1.0							
Caprylic	8	0	6.2		9.0						3.0							
Capric	10	0	3.0		8.0						50.0							
Lauric	12	0	46.5		47.0													
Myristic	14	0	21.0		20.0		0.5		0.2	0.5	15.0				2.0	7.0	6.0	
Myristoleic	14	1																
Palmitic	16	0	7.0		7.5	8.0	21.0	4.0	8.0	40.0	8.0	6.3	8.0	6.0	28.0	12.0	12.0	1.0
Palmitoleic	16	1														17.5	17.0	
Stearic	18	0	16.0	2.0	3.5	3.5	2.0	5.0	5.0	45.0	15.0	4.9	48.0	34.0	45.0	33.0	8.0	22.0
Oleic	18	1		7.0	4.0	46.0	32.0	11.0	84.6	9.0		60.6	37.0	53.0	4.0	15.0	8.0	18.0
Linoleic	18	2		4.0		41.8	44.0	50.0				21.6		2.0				2.0
Linolenic	18	3						30.0										
Ricinoleic	18	1		87.0														
Arachic	20	0																
Gadoleic	20	1				0.5	0.5		0.2			3.3	2.0	1.0				
Arachidonic	20	3§														9.0	6.0 12.0	
Behenic	22	0														9.0	6.0	1.0
Erucic	22	1																53.0
Lignoceric	24	0				0.2						2.6					7.0	2.0

* Reprinted by permission from *Emery Facts*, Emery Industries, Cincinnati, Ohio

He cites as advantages of these methods, that the raw materials are of the gas-oil type and are by-products (therefore, cheaper than paraffins); also, that they form usable by-products unlike the distillation tars produced in oxidation of paraffin.

Wittka also reported [3,4] that Germany, in 1939, produced 40,000 tons of synthetic fatty acids, which represents about 20% of the fatty acids it required for soap. The process depends on the oxidation of synthetic paraffins prepared from water-gas by the Fischer-Tropsch reaction. These hydrocarbons must be distilled so as to produce the most desirable fractions from $C_{18}H_{38}$ to $C_{28}H_{58}$, with an average molecular weight of 284. These fractions are white, very soft—i.e., more fluid than lard—and have boiling points above 350°C.

To produce fatty acids of certain molecular weight, it is necessary to start with paraffins having carbon chains about twice as long as those of the desired fatty acids.

Typical fractions employed, analyze as follows:

Carbon Atoms	*Percentage*
19–22	27.4
22–25	31.0
25–27	23.7
27–28	11.7
Over 28	1.7

The average would correspond to $C_{20.5}H_{43}$. There are four steps in the process:

1. The paraffins are preheated and oxidized with 0.1% potassium permanganate at 150°C for 10 minutes. By this means, peroxides are introduced which act as catalysts.
2. Air is blown in at 99–102°C through porous earthenware plates in the oxidation tower, thus producing a great volume of air bubbles, which completes the oxidation in 20 hours. The products of the oxidation are: carbon dioxide, volatile liquids, water and a large quantity of low fatty acids and unsaponifiable matter. The yield is about 50% of unsaponifiable matter and 50% of fatty acids.
3. The oxidized products are separated by saponifying the fatty acids and dissolving them in water. Salt is added to separate out the soap. This is split with mineral acid and the fatty acids are extracted with a solvent.
4. The fatty acids are fractionally distilled. The purified fatty acids have the following composition:

Iodine Number	4.9
Hydroxyl Number	3.7
Naphtha Insoluble	0.0

Unsaponifiable	0.29
Solidifying Point	26.2–30°C
Odor	Practically none
Acid Number	244.2
Saponification No.	247.2
Average Mol. Wt.	229.8

The mixture consists largely of lauric and myristic acids.

FATTY ALCOHOLS

There was a time when we were entirely dependent on natural sources for fatty alcohols. They were derived from coconut, palm-kernel, babassu oils, tallow, etc., by hydrogenating the fatty acids. Thus, palmitic acid is reduced by hydrogen to yield cetyl alcohol:

$$C_{15}H_{31}COOH + H_2 \rightarrow C_{16}H_{31}OH + H_2O$$

The first commercial plant in the United States, using the De Nora process for high-pressure hydrogenation of fats and fatty acids to fatty alcohols, began operations at Procter & Gamble's plant, at Long Beach, California, about 1951. This process can be conducted as a continuous or semi-continuous operation, depending on the production desired. The continuous process is employed only for more than ten tons a day. The catalyst may be either fixed, or mixed with the fat. The fixed system is employed for reactions between gases. Details of the process, an Italian invention, are given in Chapter 6. There are also several other processes.

For the past few years, alcohols C_4 and higher, have been produced by the so-called OXO process. This is a method for converting olefine hydrocarbons by a reaction with carbon monoxide and hydrogen, to aldehydes and then to alcohols. The major uses of oxo-alcohol are in plasticizers such as dioctyl phthalate, diisodecyl phthalate, diiso-octyl phthalate and tridecyl alcohol.

Until 1962, the production of straight-chain fatty alcohols was dependent on natural sources. In that year, however, Continental Oil Company started operating its Alfol alcohol plant, which promised ultimate capacity of 100 million pounds a year of alcohols C_6 to C_{20}, obtained either as high-purity single alcohols or as mixtures similar to those derived from natural sources. All this started with a license from Dr.Karl Zeigler of the Max Planck Institute, to make long-chain primary alcohols from ethylene. Other firms have also gone into the manufacture of Oxo-alcohols.

The Alfol alcohols [5] are a homologous series of straight-chain compounds containing even numbers of C atoms. They include 1-hexanol (Alfol 6) up to 1-eicosanol (Alfol 20); see Volume 2.

Alfol 12 to Alfol 20, currently find their greatest use in the manufacture of synthetic, biodegradable detergents. Three general types of surfactants, which show the desired properties as "soft" detergents, account for the major portion of total Alfol production:[6]

Alcohol sulfates—$CH_3(CH_2)_nCH_2OSO_3^-Na^+$
Alcohol ethoxylates—$CH_3(CH_2)_nCH_2CH_2O)_nCH_2CH_2OH$
Alcohol ether sulfates—
$CH_3(CH_2)_nCH_2O(CH_2CH_2O)_nCH_2CH_2OSO_3^-Na^+$

Derivatives

On account of their purity and good price range the Alfol alcohols offer broad commercial possibilities for producing a number of other useful derivatives such as alkyl chlorides, mercaptans, ethoxylates of mercaptans, nitriles, aliphatic amines, and quaternary ammonium salts.

n-Dodecyl mercaptan serves as a modifier for polymerization in the synthetic rubber industry. Ethoxylates of mercaptans function in grease-cutting, industrial cleaning, and metal-processing. The quaternaries are used as corrosion inhibitors, bactericides, emulsifiers, antistatic agents, dyeing aids, and textile softeners.

Amine oxides in detergent formulations appear to lessen skin irritation.

New Uses for Fatty Alcohols

The use of fatty alcohols as chemical retardants of evaporation may develop into large production, judging by the interest developed by important firms, who have been devoting much research to the problem and have taken out patents.

Many areas in the United States and in other countries find increasing need, not only to locate new sources of water, but also to conserve the meager supplies that they now have. This can be accomplished by spreading a micro-thin film of fatty alcohol on the surface of ponds and reservoirs.

Procter & Gamble's *Sav-A-Pond* is one such method. DuPont's Lorol 24 is used in the form of a two-pound block of cetyl alcohol in a wire cage, supported by a raft. This is sufficient for two acres of surface. Alcolac Chemical Corporation was granted U. S. Patent 3 085 850 on its process for dispersing a film of cetyl alcohol and quickly reforming the film, once it is broken by the action of waves.

Continental Oil Company's Alfol 16 (hexa-decanol-1) and Alfol 18 (octadecanol-1) derived from petroleum, may eventually be used for this purpose. Archer, Daniels, Midland Company, offer their Adol 52 and NF grade 90% cetyl alcohol.

Both cetyl and stearyl alcohols are probably most widely known for their application in cosmetics. They are used in vanishing creams, cleansing creams, shaving cream, lip sticks, and rouges. They have been used also as emollients in synthetic detergent toilet bars, when urea-fatty acid adducts are employed.

Stearyl alcohol may be used as a starting point for quaternary ammonium compounds, used as textile softeners, and in laundry softeners and anti-static agents. Stearyl pyridinium bromide is used to increase the affinity of dyes for vegetable-tanned leather.

GLYCEROL

Glycerol, or trihydroxypropane, ($CH_2OH—CHOH—CH_2OH$), is the "sweet principle of fats" discovered by Scheele, in 1783.

It is a colorless, odorless, sweet tasting, viscous, hygroscopic and substantially non-volatile liquid, with a molecular weight of 92.06. It is readily soluble in water and ethanol, insoluble in benzene, chloroform, and ether.

Anhydrous glycerol has a specific gravity of 1.262 at 20°C and boils at 290°C, with decomposition. Its index of refraction $n20/D^{1.1473}$. It is miscible with water and alcohol; soluble 1 part in 11 parts ethyl acetate and 1 part in 500 parts ether; insoluble in benzene, chloroform, petroleum ether, carbon bisulfide and oils.

Sources of Glycerol

Glycerol is a by-product in the manufacture of soap. This, with the glycerol derived from acid-splitting or hydrolysis of fats and the methyl ester process are its only natural sources.

Most oils and fats yield about 10.5%, but coconut oil yields 13.6% of glycerol.

Glycerol also is formed in the fermentation industries, but nearly all of it remains with the primary product and is not recovered.

A method invented by G. B. Bradshaw and W. C. Meuly, under USP 2 271 619 assigned to the Du Pont Company in 1942, utilizes methyl or ethyl alcohol as a substitute for the glycerol in a fat, with the resulting formation of the methyl or ethyl esters, and glycerol (see Chapter 6, for outline of process).

Applications

Industrial consumption of glycerol for 1965, has been summarized as follows:

	Million pounds[7]
Alkyds	68
Cellophane	40
Drugs and Cosmetics	60
Explosives	15
Food and Beverages	30
Tobacco	45
Polyurethane	13
Miscellaneous	41
Total	312

It has these manifold uses because of its ease of combination with other chemicals, its humectancy, or ability to absorb and retain moisture, its low volatility, excellent solvent properties, food value, etc. Its use in explosives and resins depends on its ability to form esters.

Synthetic Glycerol

There are several processes for producing synthetic glycerol, in fact, several firms have capacities, as follows:[4]

	Million pounds
Atlas	25
Dow	80
Olin (not operating 1965)	40
Shell (2 plants)	145
Total	290 per year

In 1965 the U. S. Tariff Commission and The Soap & Detergent Association reported production of 145 million pounds of natural glycerol and 201 million pounds of synthetic glycerol.

REFERENCES

1. C. R. Noller and B. A. Bannerat, *J. Am. Chem. Soc.*, 56:1563 (1934).
2. F. Wittka, *Soap, San. Chemicals*, 16(8):28; (9):34 (1940).
3. F. Wittka, *Seifensieder-Ztg.*, 67:50, 68 (1940); USP 2274632.
4. *Report of Research on Synthetic Fats at University of Leipzig:* P.B. 4291; Office of Technical Services, Washington, D. C.
5. Mark T. Atwood, *Chemistry of Alfol Alcohols;* Conoco Technical Bulletin, A 1–11.
6. Continental Oil Co., *J. Am. Oil Chemists Assoc.*, February 1963.
7. *Oil, Paint, Drug Rep.*, April 4, 1966.
8. Synthetic Glycerol: USP 2072016; 2207193; 2211626.

chapter 5

SURFACTANTS AND SURFACE ACTIVITY

The term *surfactant*, a contraction of *surface-active agent*, embraces the synthetic organic chemicals that assist in penetrating, wetting, emulsifying, dispersing, solubilizing, foaming and frothing, and finally washing and scouring. Some do not have all these properties; in others, they are more or less pronounced, according to their chemical composition.

All of these physical phenomena take place at the *interface:* that is, at the surface or area of contact between two different phases or forms of matter. Of the five types of interface—solid/solid, solid/liquid, solid/gas, liquid/liquid, liquid/gas—only the solid/liquid and liquid/liquid are of concern here.

One might point out that soap encompasses all the functions of surfactants. Soap, however, is incompatible with hard water, in which it forms curds and precipitates the calcium, magnesium, or iron. All this and more is circumvented by a good surfactant. A surfactant exhibits activity even in acidic solution. For example, a bowl cleaner (see Vol. 2) contains Triton in a solution of hydrochloric acid; another example—a floor wax—contains a fluorocarbon surfactant, in very low concentration, to improve levelling.

For many purposes, therefore, soap is not so efficient as the synthetics, so there has been a steady search for improvements.

History of Surfactants

The only natural materials with some surface-active properties were *saponins*, of vegetable origin, and *bentonite*, a clay.

In 1860, sulfonated castor oil was invented by Mercer. This is better known as Turkey Red Oil, because it was used as a wetting agent for fabrics to be dyed in "Turkey Red" with madder root, of which the active principle is alizarin. Although Turkey Red Oil helped to spread the dye evenly on the fabric it had limited detersive power.

In searching for improved detergency, Reychler, in 1913, synthesized cetylsulfonic acid and cetyl triammonium chloride. The shortage of soap in Germany during World War I, gave added stimulus to research. As a result, the first patent on a surfactant that was not derived from natural fat was filed in 1915, by Dr. Fritz Gunther, of the Badische Anilin and Soda Fabrik. It covered sodium diisopropyl naphthenate. Nekal (sodium alkylnaphthalenesulfonate) and sodium butyl naphthenate were placed on the market in 1925.

Chemical Classification

Almost all the surfactants, despite their differences in origin and methods of manufacture, have two points in common. One part, the long hydrocarbon chain that is oil-soluble (*lipophilic*), and the other portion, a water-soluble (*hydrophilic*) group. The latter property may render the whole molecule so soluble in water as to impart some or all of the characteristics mentioned.

The oil-soluble portion may be derived from a natural fat or from petroleum, such as the long, straight- or branched-chain hydrocarbon of petroleum, or aromatic hydrocarbon, such as benzene, etc., with an alkyl side chain. The water-soluble portion may be derived from mineral acids, alkylene oxides, polyhydric alcohols, and polyamines.

The surfactants are further divided into several classes: anionic, cationic, nonionic, and amphoteric. Each type acts differently in aqueous solution.*

ANIONIC COMPOUNDS

The anionic compounds form negatively charged ions, which contain the oil-soluble portion of the molecule, and the water-soluble portion carries a positive charge; e.g., sodium lauryl sulfate, $Na^+C_{12}H_{25}OSO_3^-$. Others in this group are alkylaryl sulfonates and now the "soft" linear alkylate (i.e., biodegradable) sulfonates (straight chain), which comprise about 50% of all the synthetic detergents produced. The straight-chain primary alcohol sulfates, and fatty alcohol ether sulfates also belong to this class. These constitute about 20% of all the surfactants.

* It is not the intention of this book to list all members of the various groups of surfactants. The reader is referred to McCutcheon's annual publication, *Detergents and Emulsifiers*, and other books on the subject, listed in the *Bibliography*.

An interesting type of anionic involves the reaction between the appropriate alcohol and maleic anhydride, followed by the addition of sodium bisulfite:

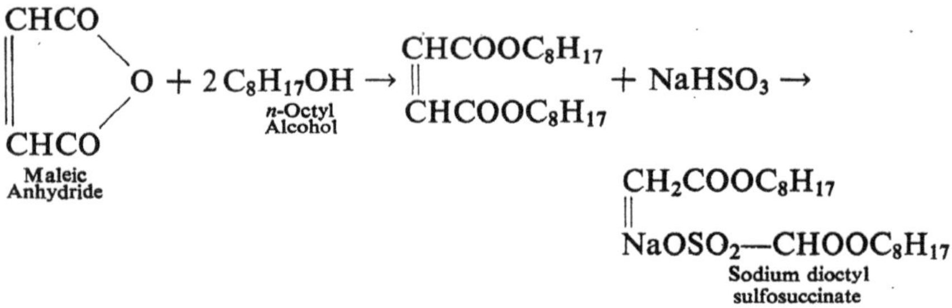

This is Cyanamid's Aerosol OT, which is claimed to be fully biodegradable.

The Igepons (General Aniline) are also members of the anionic group.

Igepon A is not stable in alkaline media, because it hydrolyzes at the ester grouping. This defect was corrected in Igepon T, in which an amide linkage, CONH, was introduced, with subsequent neutralization:

$$\underset{\text{Oleic Acid}}{CH_3(CH_2)_7CH{=}CH(CH_2)_7COOH} + \underset{\text{Taurine}}{NH_2CH_2CH_2SO_3H} \rightarrow$$

$$\underset{\text{Igepon T}}{CH_3(CH_2)_7CH{=}CH(CH_2)_7CO{-}NHCH_2CH_2SO_3Na}$$

Protein Detergents

The protein detergents,[1,2] for example, the Maypons (Stepan) and the Lamepons (Protean), which are also anionic, are condensates of fatty acids and polypeptides, produced from degradation products of protein (the collagen of leather and casein in cheese) and the higher fatty acids. As the sulfo group is absent, they are compatible with alkalies and salts. When 5% is added to soap, they reduce the sensitivity of soap to calcium. They are very mild to the skin. They are produced as coconut-protein-sodium complexes. Polypeptide 37 (Stepan) is a complex. There are also the coconut-protein-potassium and the triethanolamine products, which range from 35% to 50% active, amber liquids. They may be used as non-irritating shampoos, in oil emulsions; and Polypeptide 37 is recommended as a hair conditioner and restorer. They are among the most biodegradable products.

Sarcosinates

The sarcosinates, for example, N-acylsarcosine (Geigy), $CH_3(CH_2)_n$-$CON(CH_3)CH_2COOH$ and its salts, are anionic. These compounds may be considered as modified fatty acids in which the hydrocarbon chain is interrupted by an amido-methyl(—$COHCH_3$) group, exemplified by Maprosyl 30 (Onyx), which is a 30% active solution of sodium N-lauroyl-β-aminoacetate. They are emollients and detergents, low in toxicity and free from irritation. They inhibit formation of scum due to hardness of water.

As they are probably the least toxic anionic surfactants, they are used in dentifrices, not only because of good sudsing, but also because they have anti-enzyme potential. Moreover, they may be used as shampoos for rugs and upholstery. They produce a dry, stable foam, with little or no penetration into the backing of the rug; they dry rapidly, leaving a powdery, non-tacky residue which may be readily brushed off.

Wetting power is moderate in the alkaline *p*H range, but increases with accompanying greater foaming, and reaches its maximum at *p*H 4–5.

Phosphonols

The Phosphonols [3] (Beacon) are a series of anionic complex organic acids and polyphosphate esters. They are offered in the form of syrupy concentrates of acidic esters, at 98–100% activity. These products show remarkable wetting, emulsifying, and detergent properties, in both acid and alkaline media. They also have definite anti-corrosive properties and can sequester metallic ions. As they show a *p*H of 2–3, they should be diluted and neutralized before they are added to acid-sensitive systems, e.g., solutions of rubber latex, soaps, or alkali-soluble resins. They are soluble in water, alcohols, aromatic hydrocarbons, esters, and ethers.

Applications:

The suggested uses of these surfactants are for general purpose cleaners, for scouring textiles, cleaning up radioactive residues, dispersing pigments, polymerization of emulsions, stabilization of latex solutions, and as anti-static agents.

Oil-Soluble Sulfonates

Besides the host of water-soluble surface-active agents, there are those that are oil-soluble, e.g., those of barium, calcium, magnesium, an ashless ammonium sulfonate, 35K (Bryton); also the sodium sulfonate, which contains 0.7% of inorganic salt.

The magnesium and calcium compounds are used for oil-compounding, to introduce a high level of metal additive, in order to ensure reserve alkalinity in the finished blend. For example, Hybase M-300 (Bryton) used in lubricants for Diesel cylinders, to reduce wear, when fuels high in sulfur are used.

M-300 is a magnesium sulfonate, M.W. 920, containing 30% of active sulfonate and 47% of oil, with only 0.5% of water. In general, it might be said that the sodium salts are produced at 62% concentration in mineral oil, and with molecular weights ranging from 430 to 500. Choice for a given application is dependent on the required balance of hydrophilic *vs.* hydrophobic characteristics. If the application is for emulsification, products in the lower or middle range of molecular weight are best suited. For rust preventives and oil-system dispersion, the higher molecular weights are preferable. Calcium and barium sulfonates are primarily hydrophobic.

Uses of Sulfonated Oils

Almost every saponifiable oil is now available in its sulfated form, including tallow, sperm, and tall oil. Sulfonated oils have a variety of uses, especially in the textile, leather, and metal industries.

The oils are sold as either 75%-sulfonated—which means that they contain only 25% of water; and a 50%-sulfonated oil, which contains 50% water.

Sulfonated oils, being soluble even in hard water, lend themselves easily as adjuncts in the dyeing, wetting-out, and finishing of textile fabrics. Even though in recent years, synthetic surfactants have displaced sulfonated oils in many applications, a considerable quantity is still manufactured.

Besides their use alone, these oils may be employed to homogenize and solublize other oils and solvents, which are by themselves insoluble in water. Therefore, by the use of sulfonated oils, a great many combinations can be made to serve in the textile field as dyeing assistants, penetrants, softeners, and finishers. Compounded with mineral oils, they serve as wool oils and developers for sulfur black. They are preferable for oiling yarns because they can be easily rinsed out with water.

In the leather industry, mixtures of sulfated cod oil or sulfated neatsfoot oil with mineral and/or wool grease are used in dyeing leather, and as stuffing compounds.

Compounded with lard oil, mineral oil, rosin, fatty acids, etc., sulfated castor oil can be used as a cutting oil for metals.

Compounded with steam-distilled pine oil, sulfated castor oil forms a good disinfectant. It may also be compounded with ethylene dichloride, carbon tetrachloride, xylene, benzene, Hexalin, and Tetralin, as well as with many water-insoluble chemicals, and also with acetone and alcohol.

Sulfated castor and olive oil have been used as bases for latherless oil shampoos and hot oil treatments for the scalp. The latter use is not in such great vogue as heretofore. Oils used for this purpose must not be alkaline, but on the contrary, should contain 5–8% of free fatty acid and must be suitably perfumed. These oils are soothing to the scalp and leave the hair soft and easily manageable.

TABLE 5:1

PERCENTAGE COMPOSITION OF SULFATED OILS

	Castor	*Neatsfoot*	*Olive*	*Cod*	*Coconut*
Free Fatty Acids	13.5	31.00	14.80	23.00	12.50
Moisture	25.00	24.50	25.00	24.50	40.00
Ash	9.04	4.38	6.28	3.50	6.00
Combined SO_3	4.33	4.05	3.74	2.61	1.50
Free SO_3	0.70	0.57	0.70	0.99	0.27
Total SO_3	5.03	4.62	4.44	3.60	1.77
Neutral Oil	–	20.00	–	–	18.50
*p*H	7.00	6.20	6.65	6.35	6.50
Total Fats	65.96	70.62	69.50	71.50	54.00

Sulfated castor oil may be used to solublize perfume oils, thus obviating the use of alcohol.

Average analyses of typical sulfated oils are shown in Table 5:1.

Hydrotropes

The hydrotropes * are a series of anionic chemicals that increase the solubility of other components in a liquid detergent.[4] The Naxonates (Nease) comprise ammonium xylene- and toluenesulfonates, sodium and potassium xylenesulfonates, sodium and potassium toluenesulfonates and sodium and potassium benzenesulfonates. They are sold either as 40%-active liquids or 95% solids. Other firms manufacture them also.

They are used in the detergent field as solubilizing agents for liquid detergents containing phosphates, silicates, and other alkaline builders, in heavy duty cleaners, wax-strippers, and complex dishwashing detergents.

* Suppliers: 25, 98, 146, 165.

By incorporating a hydrotrope to the extent of 5–8% active matter, the cloud point of the system may be depressed.

A problem that often arises in formulating aqueous products is the separation of some sparingly soluble oil or solid, causing turbidity or stratification. This can sometimes be overcome by the addition of alcohols or other solvents, but sodium xylenesulfonate is often more effective as a coupling and clarifying agent. It has been found for example, that non-ionic detergents such as Ninol 1281 (an alkylolamide) or Ninox BJO (alkylphenol polyglycol ether), which are insoluble in silicate solutions, can be solubilized by the addition of a few percent of sodium xylenesulfonate.

An anionic surfactant with unusual solubility properties, for use in industrial and household applications is Dowfax 2A 1, which is the sodium salt of sulfonated dodecyl diphenyloxide. Inorganic compounds that effect good solubility are hydrochloric, sulfuric, nitric, and phosphoric acids, sodium orthosilicate, sodium pyrophosphate, sodium chloride, tetrapotassium pyrophosphate, sodium sulfate, calcium chloride, sodium hydroxide, and zinc chloride. Dowfax 2A 1 is also soluble in organic acids and organic salt solutions. 50% solutions in water can be prepared. At 77°C it is soluble and stable at 0.1% concentration in nearly all 20% aqueous solutions of inorganic compounds.

Suggested uses are: Wetting agent in concentrated sodium hypochlorite for commercial laundries, electric cleaning, wetting agent in bleaching paper and textiles, certain electroplating baths, and in chemical milling.

Dodecylbenzenesulfonates

One of the most widely used surfactants is an anionic compound of dodecylbenzenesulfonate.

Because of its relatively low price and its great utility, having profuse sudsing and cleansing properties, it forms the basis of dishwashing compounds, household cleaners, and various maintenance products.

Furthermore, the sulfonic acid intermediate lends itself to conversion to many useful products containing ammonium, sodium, potassium, and ethanolamine, without elaborate equipment and with a minimum of specialized technical personnel.

One of the shortcomings of "ABS" (alkyl benzene sulfonates) such as dodecylbenzene sulfonates however, is that they produce "hard" detergents; that is, those that are not readily decomposed by living micro-organisms. It is known that the decomposition and removal of organic substances, including detergents, from waste waters is accomplished by bacteria,—a process called biodegradation—and removal depends upon unrestricted reproduction and growth of bacteria, while using the minimum amount of the organic compounds for energy. This means that there is difficulty at the sewage disposal plants, and traces of the detergents have also been found to seep out of cesspools and septic tanks. (See section on *Importance of Biodegradability.*)

CATIONIC COMPOUNDS

The cationics ionize as *invert soaps;* that is, the oil-soluble portion is positively charged.

Quaternary Ammonium Compounds

The chief members of this group are the quaternary ammonium halides. Their outstanding characteristic is their germicidal activity, rather than detergency. Stearyl dimethylammonium chloride and others of this type are used as laundry softeners. This entire class amounts to about one per cent of the surfactant production.

Many of the "quats" are well known by their various proprietary names, for example, Hyamine 3500 (Rohm & Haas), N-alkyl (C_{14}, C_{12}, C_{16}) dimethyl benzyl ammonium chloride, the active ingredient in *General Purpose Disinfectant*, described elsewhere in the book. Others are Bionol A, (Antara), BTC (Onyx), Emulsept (Witco), Germ-i-tol (Fine Organics), Roccal (Sterwin), Zephiran (Winthrop) and many others. They are used primarily as disinfectants, germicides, fungicides.

Those based on stearic acid or tallow are used also as hair rinses, textile and laundry softeners and anti-statics; for example, Pendit CA (Rayette, Inc.), is used as a rinse for human hair to which it is substantive. It is sold as a 25% paste, stearyl dimethylbenzylammonium chloride. Another such is Ammonyx 4, (Onyx). Ammonyx 2194, ditallow dimethylbenzylammonium chloride is used in softeners for household laundering and for paper and textiles.

Amine Compounds

To this class also belong the primary and secondary amine salts and acylated polyamines. The series of tertiary amine oxides also are members of this class: Examples are:

Ammonyx

LO:	30% lauryl dimethylamine oxide		
CO:	30% cetyl	,,	,,
MO:	30% myristyl	,,	,,
MCO:	30% ,, -cetyl	,,	,,
SO:	30% stearyl	,,	,,
TO:	30% Tallow	,,	,,

These amine oxides, though they are all cationic, are compatible in all proportions with sodium lauryl sulfate and the ether sulfates; also the biodegradable homolog of sodium dodecylbenzenesulfonate (Ultrawet—a product of 40% solids). These amine oxides are compatible with the amphoterics, such as Maprolyte 88 and 92, and with Onyx BTC 2125, which is a quaternary. The amine oxides are biodegradable also. They may be usefully applied in detergents, foam stabilizers, polymerization inhibitors, initiators, components of polymers, propellant promoters, light stabilizers, textile assistants, petroleum additives, electroplating additives, and many more uses.

Onyx Chemical Corporation reported that in a home-use panel test conducted with the cooperation of twenty-two women, eighteen were able to detect the difference in the feeling of the amine-containing detergent, as compared with another sample used as control, in which no amine oxide was present. They preferred the emolliency that the amine oxide apparently contributed. The superior emolliency was observed by the same women in an experimental formula, in which Ammonyx SO and CO were included.

NONIONIC COMPOUNDS

The *nonionics* do not ionize, but acquire their hydrophilic character from an oxygenated side chain, usually polyoxyethylene. The oil-soluble portion may be derived from fatty acids, alcohols, amides, or minaes. This type comprises about 25% of all detergents. They have the additional property that they may be combined for use with either anionic or cationic surfactants. Examples are the Igepals, Tritons, Tergitols, Surfonics, Plurafacs.

The members of this class are of varied composition. For example, Igepal CO630 is 100%-active nonyl phenoxypolyoxyethylene ethanol. Neutronyx 600, same as the first, as are the Hyponic PE series of Nopco. Ninox BJO (Stepan) alkylphenol polyglycol ether; the Pluronics (Wyandotte) are 100%-active condensates of ethylene oxide with bases formed by varying the ratio of ethylene oxide to the hydrophobic (water-insoluble) portion of the base, compounds are formed which yield more or less suds. Poly-Tergents (Olin) are aromatic and also aliphatic polyglycol ethers. Triton X100 (Rohm & Haas) is isooctyl polyoxyethanol. As a rule, none of these produce copious suds. They are good cleaners and emulsifiers and are useful as components for machine dishwashing and in rinses for glassware and other hard surfaces.

Ethylene Oxide Condensates

Condensates of ethylene oxide with alkylated phenols can be produced with a variety of properties. The alkyl phenols (hydrophobic) are very soluble in oils, whereas ethylene oxide is infinitely water-soluble (hydrophilic). They can be combined in almost limitless hydrophobic-hydrophilic ratios. Depending on the number of moles of ethylene oxide in the compound, products may be prepared that satisfy almost every purpose for which a surfactant is used, such as defoamers, emulsifiers, detergents, and wetting agents. The structural formulas of these products may be represented as follows:

$$C_8H_{17}\text{—}(OCH_2CH_2)_xOH$$

x represents the number of moles of ethylene oxide.

Two series of nonionic surfactants, in order of increasing length of the polyoxyethylene chain, are shown in Table 5:2

TABLE 5:2

TYPICAL ETHYLENE OXIDE DETERGENTS*

Octyl phenol series	*Nonylphenol series*	*Ethylene Oxide groups*
Tritons:		
X-15		1
X-35		3
X-45	Tritons:	5
X-114	N-57	7–8
X-100	N-101	9–10
X-102	N-128	12–13
X-165		16
X-205		20
X-305		30

* Rohm and Haas.

These surfactants (type of Tritons) are used in industrial and household detergents in liquid, paste, and powdered cleaning compounds ranging from heavy-duty industrial to gentle detergents for fine fabrics. They may be blended with acids, alkalies, excipients, and water softeners for any desired characteristics.[5]

The viscosity of solutions of members of this series increases on approaching a content of 50% from either higher or lower concentration. The use of some of these for various applications are described elsewhere in the book.

Fatty Acid Alkanolamides

Another group of surfactants, also non-ionics, are the fatty acid alkanolamides. They are produced by the so-called Kritchevsky reaction, whereby fatty acids are condensed with diethanolamine.[6]

These products stabilize the foam of alkylaryl sulfonates and ensure good cutting of grease. They are used in high-sudsing detergents, to boost the foam. Besides these properties, they increase viscosity in detergents and shampoos. They combine good cleansing and rust-inhibiting powers. They are useful in dispersing resins and latexes. How they are used, is shown in Volume 2.

Sugar Ester Surfactants

The sugar ester surfactants are novel nonionic compounds, produced under basic U.S. patent 2 893 990 to Sugar Research Foundation. The reaction involves alcoholysis of sucrose with methyl esters of fatty acids, using potassium carbonate as catalyst and dimethyl formamide as solvent. It is believed that esters can be produced to contain as little as 50 ppm of dimethyl formamide degradation products.[7,8]

An important feature of these surfactants is that they are biodegradable. Sucro-chemical Division of Colonial Sugars Company, Gramercy, La. offers mono-, di-, and triesters of tall oil, tallow, and oleic acid. Japanese investigators have reported esters derived from ricinoleic acid and 12-keto-oleic acid. Dicarboxylic acids have been used to esterify sucrose.

Monsanto has patent 2 868 781 which it has not currently been used. Procter & Gamble also holds patents for sucro-esters, using dimethyl sulfoxide, dimethyl acetamide, and other homologs as solvent. Ledoga of Italy, licensed under Sugar Research Foundation patents, is experimenting toward sugar-ester surfactants by direct treatment of natural fats.

The stiff consistency of sucrose monoesters makes them especially adaptable for use in detergent compositions that are to be used as powders, granules, and bars.

The diesters of sucrose range from high-melting waxes to thick oils, and these varied properties should be of value in formulating polishes, lubricants, metal-working compounds, paper coatings and impregnants, textile specialties, and coatings for modifying the surfaces of food products.

The sugar ester surfactants are relatively nontoxic and nonirritating to the skin. They should be of interest to the pharmaceutical profession, and as detergents for washing fruit, vegetables, and food-handling equipment.

In detergent formulations with the usual alkalies and adjuvants, sucrose monoesters of tall oil fatty acid, sucrose monotalloate, and sucrose mono oleate, were shown to remove as much, or even more, soil than similar formulations with alkylbenzenesulfonate and tall oil polyethylene glycol ester.

AMPHOTERIC SURFACTANTS

In addition to the well-defined types of surfactants, there are many that fall into the class of *amphoterics;* that is, they act as both cationic and anionic compounds. Of the very large numbers of different amphoteric surfactant compounds and variations, the products of only five manufacturers will be described here as representative types.

Deriphats (*General Mills*) are amphoteric surfactants that contain both carboxylic and amino functions in their structure. Like others of their class, they are strongly cationic in acid media and anionic in alkaline solutions. In the range of "normal use"—*p*H 5 to 9—the Deriphats can exhibit both cationic and anionic behavior simultaneously.

Deriphats 151, 154, 160C and 170C, may be described as N-substituted aminoacid derivatives (β-alanines). They are produced by condensing fatty primary amines with acrylic monomers. The reaction can be controlled to yield two types of products:

1) The *mono-adduct*, $RNHCH_2CH_2COOM$;* and
2) The *di-adduct*, $RN(CH_2CH_2COOM)_2$

Deriphats 151 and 170 C are typical balanced amphoteric surfactants containing one amine and one carboxylic functionality. Deriphats 154 and 160 C are unbalanced amphoterics, containing two carboxylic acid groups and one amino group. These products are

* M represents salt-forming cation or hydrogen.

versatile and function over a wide range of *p*H, either strongly acid or strongly alkaline.

Detergency exhibited by the Deriphat molecule stems from its structural relationship to soap (straight-chain amino carboxylic amphoterics are sometimes called "interrupted soaps"). The presence of the amino group provides excellent tolerance to hard water. Detergency falls to the minimum at the isoelectric point.* Hence it is possible to obtain almost any desired degree of detergency, e.g., controlled cleaning of the hair and scalp, without excessive defatting. These compounds are substantive to hair and skin. This property is valuable also in textile applications, such as cleaning and scouring wool, lubricating, and as dyeing assistants, via controlled *p*H.

Applications are in shampoos, detergent bars (BP 745 367 Unilever), emulsion polymerization (USP 2 688 609, to Hercules), oven cleaners, electroplating, light- and heavy-duty household detergents, emulsifiers in latex paints, corrosion inhibitors.

Antaron F C-34 (Antara) is a 37%-active complex fatty amido compound (the fatty group being derived from coconut oil). It is amphoteric and supplied at *p*H 12–13. Aqueous solutions will, without separation, tolerate concentrations of 10% or more of sodium carbonate, sodium sulfate, phosphates, and silicates. It is stable also to 10% solutions of sulfuric acid, and fairly stable in hard water.

It is well suited for the simultaneous fulling and neutralizing of carbonized woolens.

Miranols are a representative type of amphoterics based on USP 2 528 378 and others. Their generic chemical structure may be represented as follows:

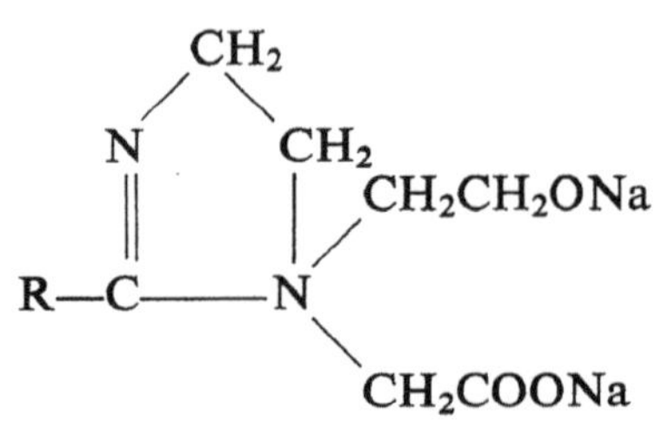

R represents the fatty acid radical and therefore it is different for each derivative. It can be lauric, myristic, stearic, octoic, etc. Chem-

* For each aminoacid, there is a definite hydrogen-ion concentration at which the acidic and basic ionization is equal. This particular *p*H is known as the *isoelectric point.*

ically, these surfactants may be described as alkyl (fatty acid) ethyl cycloimidinium, 1-hydroxy-3-ethyl sodium alcoholate-2 methyl sodium carboxylate. They are good foamers and stable over the whole *p*H range. They have some germicidal, germistatic, fungicidal, and fungistatic properties. They are nontoxic and do not dry or irritate the skin. Their uses vary from shampoo base for babies to heavy-duty cleaners for floors, dishwashing machine compounds, and general industrial cleaners.

Uniterge amphoteric surfactants (Universal Chemical Corp.) are either water-soluble or oil-soluble compounds. The water-solubles are also soluble in methyl, ethyl and isopropyl alcohols. At the isoelectric point, they become soluble in most organic solvents and there is a significant partition of the hybrid ion into the layer of solvent.

The oil-soluble Uniterges BC and BO are dispersible in water and soluble in acidic aqueous media and in alkaline solutions. They are soluble in alcohols, ketones, ethers, aliphatic and aromatic compounds, and thus have proved suitable for use in processing metals, as corrosion inhibitors, and as petroleum additives.

Because of their proteinaceous structure they act as protective colloids. They also have application in cosmetic, textile, leather, paint, dry cleaning, anti-static, and household sanitizers.

Triton Q S-15 (Rohm & Haas) is a liquid, anhydrous, amphoteric surfactant. It is a water-soluble, 100%-active, oxyethylated sodium salt containing both anionic and cationic centers. It exhibits excellent detergency, solubility, and stability in aqueous solutions of strong alkali (including caustic), and is particularly recommended as an ingredient in highly alkaline built cleaners, where maximum detergency is required.[9]

It is also soluble in all proportions in alcohols, aromatic solvents, chlorinated solvents, Cellosolve, glycol ethers, and ketones. It is insoluble in alphatic solvents such as heptane, VMP naphtha, kerosine, and mineral oil.

Triton Q S-15 is compatible with both anionic and nonionic surfactants.

When added to dry alkaline builders, some discoloration may develop, but there is no loss in surface activity.

CAUTION: Like most organic compounds, it should not be used or compounded with strong oxidizing or reducing agents, because mixtures of these materials with most organic compounds may be explosive.

Efficient Surfactants

The cleaning efficiency of various surfactants was measured by the following procedure: Mild steel panels coated with petroleum were cleaned by slowly rotating them for five minutes in a 5% solution of a highly alkaline cleaner at 85°C. The alkaline composition was:

Caustic soda	30
Sodium metasilicate	30
Soda Ash	35
Surfactant	0–5

Residual oil was estimated by observing the water-break pattern, after spraying with cold water. In Table 5:3 are given data comparing the effectiveness of Triton QS-15, and several surfactants now used in heavy-duty alkaline cleaners.

TABLE 5:3

EFFECT OF SURFACTANT ON EFFICIENCY OF ALKALINE CLEANERS

(Soil: SAE 40 Oil)

Surfactant in Alkaline Blend	*Efficiency of Alkaline Cleaner, %* Active Surfactant in Alkaline Blends		
	1%	2%	5%
None	0	–	–
Triton QS-15 [1]	99	100	100
Sodium lauryl sulfate	–	74	100
Sodium alkyl aryl sulfonate	39	73	98
Triton X 100 [2]	91	98	100
Rosin soap [2]	–	–	49
Fatty amido complex	23	–	–
Coco imidazolinium dicarboxylate	–	51	–
Coco aminopropionate	–	–	22
Isopropyl naphthalene sulfonate	–	0	95

1 Triton QS-15 is not outstanding under acid conditions. Properties dependent upon surface activity are optimum at high *p*H.

2 Not completely soluble at 85°C in alkaline solution.

IMPORTANCE OF BIODEGRADABILITY

A discussion of linear alkylate sulfonate or "soft detergent" would not be complete without mentioning a little of its historical background.

Several years ago, when the general use of detergents became almost world-wide, West Germany appears to have sensed the shortcomings of branched-chain alkylate benzenesulfonate, because it appeared to be passing through the sewage disposal plants and the suds later appeared in rivers and finally in drinking water. A law was passed prohibiting the use of surfactants that did not decompose at the sewage disposal plant or soon after the effluent reached the river.

The same phenomenon was noticed in Long Island. The daily newspapers soon spread the alarm and Congress began to stir in the direction taken by West Germany.

To ward off restrictive legislation concerning contamination of water supplies by "hard" detergents, the soap and detergent manufacturers, of their own volition, promised to have ready for distribution, their "soft" biodegradable detergents, by June 1965.

Biodegradation is the bacterial breakdown of the organic molecule after it has passed through the sewage disposal system and out into the river. Those that decompose in a matter of a few days are called *biodegradable*.

It is known that decomposition and removal or organics including detergents from waste waters is accomplished by living microorganisms, and that depends on unrestricted reproduction and growth of bacteria, while using a minimum amount of the organics for energy.

Authorities claim that pollution by detergents is only a minor part of the over-all problem. Much of the sudsing in the water is due to other materials, and despite the change in surfactants sudsing will continue because of faulty sewage disposal systems and effluent from cesspools and septic tanks. Consequently millions more must be spent before sudsing of water can be alleviated.

BIODEGRADABLE SURFACTANTS

Spokesmen for the industry say that about 500 million pounds of the linear alkyl sulfonates will be available to replace the old "hard" ABS ingredients to produce 4 billion pounds of finished products a year.

Some manufacturers of either linear alkylate benzene sulfonate (LAS), linear alkyl sulfates, or ethoxylates say that their products degrade from 85–95% in a matter of days. Surfactants based on natural organic polyhydrols and fatty alcohols are said to biodegrade completely ın about two days, under favorable conditions.

Surfactants based on organic polyhydrols and fatty alcohols are said to decompose completely in about two days, under favorable conditions.

The battle against stream pollution still remains and will be mitigated only by improved facilities for treatment of sewage.

Authorities say that the degree of biodegradability achieved by the new detergents, is greater than that legally required by the West German government. Improperly designed sewage treatment plants will not remove detergents or other pollutants. It is generally agreed that the presence of foam in drinking water indicates pollution that goes beyond that of detergents alone.

In 1965, more than 60 million people lived in unsewered areas and the sewage contributed by most of the 57 million people living in sewered areas received less than secondary treatment, so that there is still much to be done towards mitigation of water pollution.

The linear alkylate sulfonate (LAS) type of surfactants are more resistant to bacteria under anaerobic conditions, such as septic tanks and cesspools, than was at first expected. Also, that the degree of degradation for LAS under some laboratory tests is not fulfilled under the often less than optimal conditions in municipal sewage plants is also recognized. Researchers are working assiduously to correct these faults, and it is expected that more degradable detergents will be developed. Among the latter may well be the ethoxylates and other derivatives of primary, secondary, and natural alcohols, alkane sulfonates, sugar esters, and those of the alpha-sulfo fatty acids.

The industry is in such a state of flux that all the biodegradable products have not yet been included in McCutcheon's 1965 Annual. What follows is merely a sampling of the surfactants mentioned in the text.

As for the alkylates, Atlantic Refining Co., Continental Oil Co., Esso Research & Engineering Co., Gulf Oil Corporation, Union Carbide and others offer biodegradable products.

A letter from Alcolac Chemical Corporation Oct. 1965, announced that they have 29 different biodegradable surfactants. They are sold under the trade names—Abex, Sipex, Sipon, Siponate, and Siponic. They include anionic, nonionic, and cationic surface-active agents, employed throughout the chemical and related industries. They are used primarily in cosmetics, toiletries, pharmaceuticals, paints, plastics, textiles, and household products.

Among other new or modified products offered to date are:[10]

ANTARA (GAF): Several Igepals—dodecyl phenol ethylene oxide condensates.

ARDMORE CHEMICAL CO.: Ardets—liquid and powdered anionic compounds.

ARMOUR INDUSTRIAL CHEMICAL CO.: Aromox line—amine oxides;

recommended for shampoos and cleaners, and useful in many other industries.

ATLANTIC REFINING CO.: Many soft Ultrawets, anionic and nonionic.

JEFFERSON CHEMICAL CO.: Surfonics—nonionic compounds, for light-duty detergents.

MARCHON, LTD.: Empicol TAS series of products.

MIRANOL CHEMICAL CO.: All amphoteric surfactants now 90–92% biodegradable.

MONSANTO CHEMICAL CO.: Many soft alkylates, DDBSA compounds, and Santomerse formulas; anionic and nonionic compounds, for light- and heavy-duty detergents.

NATIONAL ANILINE DIV. (ALLIED CHEMICAL CORP.) claims Nacconols now are a complete line of LAS-based detergents; liquid, flake, spray-dried products.

NOPCO CHEMICAL CO.: Hyonic LA and FA series; all completely biodegradable.

OLIN CHEMICALS DIV.: Poly-Tergents—100% nonionic liquid alcohol alkoxylates.

PETROCHEMICALS CO. offer some low-foaming, quickly degraded, anionic products; one is especialy good in electroplating.

PILOT CHEMICAL CO. offer Calsoft flakes and liquids—high-sudsing biodegradable products.

ROHM AND HAAS offer several low-foaming Tritons for a diversity of uses.

STANDARD CHEMICAL PRODUCTS, INC. feature Standafome for bubble baths and cosmetic preparations.

STEPAN CHEMICAL CO offer new biodegradable products under all their brand names; Ninols, Maypons, Stepanols, Biodets, etc.

SWIFT AND CO.: Solar surfactants are available as biodegradable products—anionic and nonionic.

SURFACT-CO, INC. offers their Surco line for all-purpose synthetic cleaners.

ULTRA DIV. (WITCO CHEMICAL CO.) offer Sulframins, for specialty cleaners, emulsifiers, etc.

UNION CARBIDE offers their Tergitols, both anionic and nonionic surfactants, ranging from oil-soluble to water-soluble.

WYANDOTTE CHEMICAL CORP. claims that their Plurafacs are 100% active; available as liquid, paste, and solids, designed to meet all criteria.

SEWAGE WATER AND B.O.D.

The effluent from sewage must not be discharged into a river until it is practically neutral; that is, its *p*H should be between 5 and 9. Furthermore, it must not be discharged into the river with so high deficiency of oxygen, as to take up the dissolved oxygen from the river, thus causing death of fish and the development of putrid fermentation on the river bed. Such action could contaminate an entire area and sometimes even the neighboring underground strata.

The putridity of water is evaluated by the measurement of the *biochemical oxygen demand* (BOD).[12]

It is obvious that if effluent water is discharged into a large and fast-flowing river, a lesser degree of purification is necessary than if the same water were discharged into a small, slow-flowing stream, as the latter carries much less oxygen than the large river. The oxygen is necessary to facilitate the self-purifying process which takes place within the water course. Furthermore, the BOD measurement gives an indication of the time that will be necessary for the biological purification and of the quantities of air to be used.

Although some members of each class of surface-active agents have been cited, many have been omitted, such as the alkylene mono- and distearates and the various other fatty acid esters, which are used in detergents, as thixothropic, pearlescent, opacifying, and emollient agents. To accomplish the foregoing purposes, they are normally used in low percentages in the formulations that are to follow in Volume 2.

ANTI-POLLUTION LAW

NOTE: As a result of President Johnson's signing the *Water Quality Act of 1965*, on October 2, 1965, the Secretary of Health, Education and Welfare, is given authority to stop any individual, industry or local government from polluting water.

The effect of the new law, will be the upgrading of many sewage-treatment plants.

When discussing sewage-treatment plants, mention is made of primary and secondary treatments.

Primary plants remove about 45% of the solids and 35% of the oxygen-consuming bacteria through settling basins.

Secondary plants remove about 85% of solid waste and oxygen-consuming bacteria before the effluent is run into the river.

REFERENCES

1. H. L. Sanders and M. Nassau, Protein-based Detergents, *Soap Chem. Specialties*, Jan. 1960, p. 57.
2. R. R. Riso, Protein-derived Detergents, *Ibid.*, May 1963.
3. R. L. Mayham and F. Krupin, Phosphate Surfactants: Properties, Uses, *Ibid.*, April and May 1962.
4. M. Mauser and P. Sosis, Hydrotopes in Liquid Syndets, *Ibid.*, Feb. 1962.
5. R. Getty and W. Stericher, Liquid Detergents, *Ibid.*, Oct. 1960; June 1961; Jan. 1962; May 1962.
6. Kritschevsky, USP 2 089 212, 2 094 608/9, 2 096 749, 2 173 058; (to Ninol Laboratories, now Stepan Chemical Co.).
7. H. B. Haas, Sugar Ester Surfactants, *Manufacturing Chemist*, 29:152 (1958).
8. Osipow, Snell, York, Finchler, Sugar Ester Surfactants, *Ind. Eng. Chem.* 48:1459 (1956).
9. D. B. Fordyce, R. E. Wolfrom, and Jean Dupré, Built Nonionic Liquid Detergents, *Ibid.*, (1) Mar. 1962; (2) Apr. 1962.
10. Summary of articles in *Ind. Eng. Chem.*, September, 1964; *Oil, Paint, Drug Reptr.*, May 24, 1965; personal communications.
11. Testing Procedure of Soap and Detergent Association, *Soap, Chem. Specialties*, April, 1965.
12. *Water Treatment Handbook* (1st Eng. Ed.); Ets. Emile Degremont, Suresnes, France, 1955.

Supplementary Reading

Advances in Soaps and Detergents, *J. Am. Oil Chemists Soc.*, November 1963:
- M.M. Cohn, The Nation's Water Pollution Problem . . . , 645.
- H. D. Swisher, The Chemistry of Surfactant Biodegradability, 648.
- W. A. Kline, The Effects of Detergents on Surface and Ground Water Problems, 656.
- C. A. Houston, Foreign Requirements and Developments in Biodegradability, 659.

American Society of Civil Engineers, *Report of Detergent Problems in Perspective*; April 1965 (see *Bibliography*).

Doris Clark, Biodegradable Surfactants, *Soap Chem. Specialties*, August 1965, 57.

A. Eldib, Biodegradability of Amphoteric Detergents, *Ibid.*, May 1965, 77.

D. J. Foster and R. R. Fields, Synthetic Detergents Based on LAS, *Ibid.*, August 1964, 24.

R. L. Huddleston and E. A. Stetzkorn, Biodegradability of Detergent Hydrotopes, *Ibid.*, 1965, 57.

E. A. Knaggs, Alkylolamides in Soft Detergents, *Ibid.*, December 1964; January 1965.

P. V. Knapp *et al.*, Field Study of the Removal of LAS Detergents by the Activated Sludge Process, *Ibid.*, July 1965.

J. C. Lockwood, The Detergent Industry and Clean Waters, *Ibid.*, December 1965.

C. J. Stanberry, Jr., Performance of LAS Syndets, *Ibid.*, March 1964.

W. A. Sweeney, Biodegradability of Linear ABS, *Ibid.*, March 1964.

L. D. Berger, Nonionic Surfactant Activity, *Ibid.*, March 1961.

chapter 6

PRODUCTION OF FATS AND OILS

Most of the vegetable oils are produced from beans, seeds, or nuts. In general, the manufacturing process consists of freeing the seed, bean, or nut from lint, burrs, or shell; drying, crushing, cooking, and then pressing by hydraulic pressure, or by other means, so as to separate the oil from the meal.

In the South, cottonseed is crushed as closely as possible to the source of the seeds. The crude oil is then shipped for further processing to northern refineries, where there is a greater demand for both oil and by-products.

Expeller Process

In the expeller process, which is employed to a great extent because it requires less labor than pressing, the seed passes through a sort of meat-chopper, and both the oil and cake are expelled by machine (Figure 6:1).

Solvent Extraction

By another method, the seed is ground and the oil is extracted with a volatile solvent, which is then distilled off, leaving the oil, which is later refined and deodorized. The outstanding advantages of one of the latest extraction processes, in which hexane is employed as a solvent, are the greater yield of oil and the greatly increased protein content of the meal. In soybean extraction, one manufacturer claims that by this process only 1% of oil remains in the meal.

The solvent extraction process used by A. E. Staley Manufacturing Company, includes cracking and flaking the beans into particles of

FIG. 6:1. ANDERSON DUO-EXPELLER FOR EXTRACTING OIL FROM SEEDS
(*V. D. Anderson Co., Cleveland. Ohio*)

uniform size. The flakes are carried by a belt-conveyor into another building, where they are emptied into baskets which are moved to the top of an extraction tower. The baskets are perforated at the bottom and the solvent, which is introduced at the top of the tower, drops from one basket to another. The oil-laden solvent is drained from the bottom of the tower and the oil-free flakes are carried away for removal of solvent by drying and steaming. The solvent is distilled off and the oil, freed of solvent, is ready for commerce as *crude*, or is refined by further processing.

Filtration-Extraction of Cottonseed Oils

For every pound of cotton produced, two pounds of cottonseed becomes available, which indicates the importance of the seed. Cottonseed oil is normally worth four to seven times the price of the meal. It is natural, therefore, that the industry should adopt solvent extraction because of its larger yield of oil.

Direct application of hydraulic pressure was one of the earliest methods used. To-day about 35–45% of the crop is processed in screw presses. The first modern continuous installation for solvent extraction of oil was completed in 1947, at Wilson, Arkansas, by Delta Products Company. But only 7–8% of the 1950–51 crop was processed by solvent methods. By 1957, twenty-eight mills used solvent extraction, of which eleven have direct processes. Their installed capacity of about 7000 tons a day represents approximately 40% of the crop being processed.

The average oil content of hydraulic press-cake ranges from 4–5 percent. The residual oil in screw press-cake is 3–4 percent. Compared with hydraulic pressing, the principal advantages of screw-pressing, are higher yield of oil, greater capacity, and lower costs for labor. It has the disadvantages of increased consumption of power and high costs of installation and maintenance.

Direct solvent extraction has been developed to an efficient operation producing a residual content in the extracted meal of only about one per cent of oil. Solvent extraction plants are costly, however, and require well-trained operators to maintain the efficiency.

In 1949, pre-pressing ahead of solvent extraction was introduced. It has been found that this simplifies the operation to an extent, and further increases the yield of oil. Less solvent is lost, safety is improved. The rate of extraction of better formed flakes is increased greatly. All operating costs except power are reduced. Pre-press solvent extraction recovers more oil than other processes. Most of the oil is removed by the presses, leaving about ten per cent in the cake to be removed by solvent extraction.

Filtration-extraction differs from direct and pre-press extraction, in that it includes a slurrying step (temperature 125–135°F) and 30 to 45 minutes to replace the usual counter current diffusion-extraction employed by the other processes.

In 1957, the filtration-extraction plants cost from $300,000 upward for a unit processing 100 tons of cottonseed a day. Such a plant was engineered by Wurster and Sanger at Hollandale, Miss. The basic patents have been assigned to the Secretary of Agriculture.[1]

Further work was done at the Southern Regional Research Laboratory of the U. S. D. A.;[2] and through active co-operation between the Lukenweld Division of Lukens Steel Company, and Mississippi Cottonseed Products Company.

Fig. 6:2 is a flow sheet for the filtration-extraction process.[3]

Alkali Refining

Crude oil, as it is received at the refinery, regardless of the process by which it has been won from the seed or other source, contains objectionable impurities. These may consist of pulp or fiber, sus-

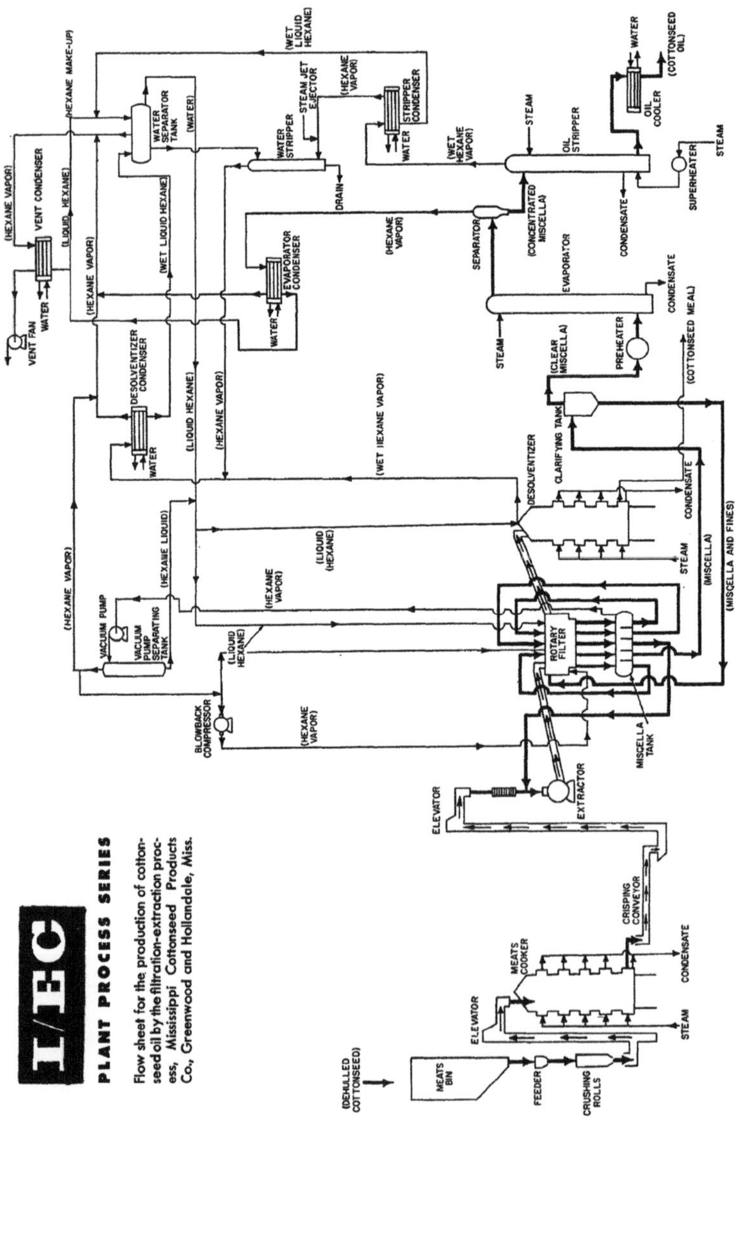

FIG. 6:2. FILTRATION-EXTRACTION PROCESS
(Reprinted with permission of *Journal of Industrial and Engineering Chemistry*)

pended albuminoids, phosphatides, sterols, and other components of the seeds. During storage and handling, there is usually a slight decomposition of the oil, which is evidenced by the presence of free fatty acids. The extent of decomposition depends on the time which has elapsed since the gathering of the seeds or nuts. Decomposition consists of a gradual breakdown of the glycerides.

Crude oil may contain aldehydes, ketones, lactones, oxy- and hydroxy-fatty acids, acrolein, carbon dioxide, and moisture. It is believed that the volatile aldehydes and acids are chiefly responsible for rancidity. Moreover, traces of metallic substances as well as enzymes from the seed may lead to formation of aldehydes and other impurities that cause rancidity. The refining and deodorizing processes remove all these impurities.

In refining an oil, the first step is to determine the content of free fatty acids of the sample by titration with standard alkali. From the result can be calculated the exact amount of caustic soda that will be required to neutralize the free fatty acids in the batch. For example, if it is determined that the crude oil contains 0.75% free fatty acids, then 1000 lb oil contains 7.5 lb of free fatty acids that are to be neutralized.

A 20°Bé solution of caustic soda is generally employed for refining. Such a solution contains 14.3% of pure caustic soda. Furthermore, it has been ascertained that 140 lb of pure caustic soda is necessary to neutralize 1000 lb of fatty acids. Therefore,

$$7.5 \times 0.140 = 1.05 \text{ lb of } 100\% \text{ caustic soda}$$

will neutralize 1000 lb of oil.

When employing a 14.3% solution,

$$\frac{1.05}{0.143} = 7.3 \text{ lb of } 20°\text{Bé solution}$$

will be theoretically required for refining 1000 lb of the oil.

In practice, however, usually slightly more than the theoretical amount is necessary. The excess lye to be used may vary from 0.1% for relatively clean oil, to 0.5% for a gummy oil.

Laboratory tests are usually conducted to determine the loss in refining and the color of the finished oil. The loss amounts to 1–3% minimum to a 7% maximum, depending on the character of the crude and whether it contains more or less surface-active material, which tends to emulsify and occlude neutral oil. Thick, gummy crude cottonseed oil, for example, may show a refining loss of 7 per cent.

The refining kettle is a cylindrical steel tank with a conical bottom. It is fitted with a bottom valve for drawing off the "foots" (residue) and with a swinging draw-off pipe for drawing off the refined oil. It is usually equipped with both open and closed steam coils and an agitator. The oil is heated to 90°F and, during stirring, the requisite amount of lye is added from an overhead tank, by means of a sprinkler, so that the lye will be evenly distributed over the surface of the oil. A common practice in refineries is to operate on a tank car of oil as the batch; large refineries, however, may work with five or six cars at a time, in a single refining tank.

Several minutes after the lye has been sprinkled into the batch, brownish or dark clots form in the oil while it is being agitated. These clots consist of saponified fatty acids, together with gelatinous cellular impurities. A sample drawn in a bottle, shows that the dark clots are intermingled with bright yellow oil. The oil is now heated to 120°F and the agitation is stopped. Then the batch is allowed to settle for five hours or over night. The supernatant oil is pumped off to the bleaching kettle by gradually lowering the swinging draw-off pipe from the top to the surface of the soapstock or "foots."

Centrifugal Refining Machine

There are several methods for the continuous refining of oil.[4] In one of these, the alkali (caustic soda and/or soda ash solution) is introduced by means of proportioning pumps which feed the two ingredients in the proper proportion. They are intimately mixed, heated to 250°F to break the emulsion, and then run through a battery of centrifugal separators. This method is valuable because of the speed of operation, larger yield of clear oil, and often better color. The oil is washed by mixing with hot water and again centrifuged to remove residual soapstock. Finally, it is sprayed into a vacuum dehydrator to flash off the remaining traces of water. It should be borne in mind that oils that are to be stored must be absolutely dry, or decomposition soon sets in.

The Podbielniac Duozon and Hydrozon centrifugal refining machines for vegetable oils, introduced in 1956, are said by the manufacturer to have demonstrated superior performance as compared to conventional high-speed centrifuges, in continuous processes for degumming, alkali-refining, and water-washing of a variety of oils.

Figures 6:3 and 6:4 show the essential components of the machine used for single-step refining and washing-water operations. These machines are 36 inches in diameter and have capacities up to 30,000 lb an hour.

In refining crude coconut oil, the oil and the caustic soda solution of required strength are pumped under proportional control, heated to 160–180°F, and then blended continuously in a short section of pipe line, before entering the machine. The vigorous mixing which is customary in the refining of most other oils, is avoided here because of the tendency of coconut oil-soap mixture to form stable emulsions which make subsequent separation difficult.

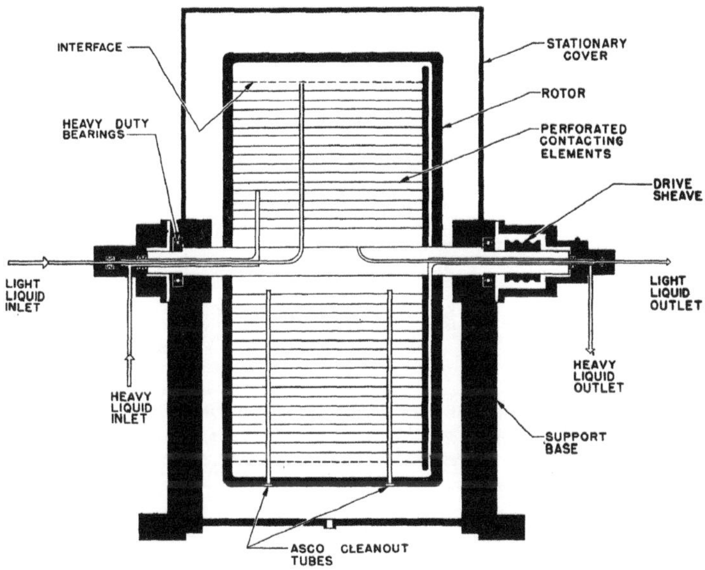

FIG. 6:3. CENTRIFUGAL MACHINE FOR REFINING VEGETABLE OILS (*Podbielniak Div., Dresser Industries, Franklin Park, Ill.*)

The soap-oil mixture is then introduced into the rotor, at a point near the periphery where the heavier soap-stock phase is separated. Hot wash water (10–15%) pumped into the rotor, flows countercurrent to the flow of the separated oil, and combines with the soap-stock phase to be discharged in a single stream. The interphace position of the heavy and light phases, within the rotor, is automatically controlled, as previously described, by the back pressure valve in the discharge line.

Ratios in flow of oil and caustic are controlled by employing a rotary positive-displacement pump, with a variable speed drive to pump the crude oil at the desired flow rate. Caustic is introduced with positive-displacement piston pump, provided with manual stroke adjustment. A sample of the mixture may be withdrawn for analysis of total alkalinity, and if necessary, the flow of caustic is reset, to provide the proper ratio. Accuracy of total alkalinity in the mixture can be maintained within ±2 percent.

The filtered crude oil is refined and washed with water, after which the residual color and soap are removed by vacuum-bleaching with 0.1–0.2% of fuller's earth. A high-vacuum decolorization, produces oil light in color, with practically zero soap content. The oil is given a final "polishing" filtration before being bottled for sale (this installation is in Panama City).

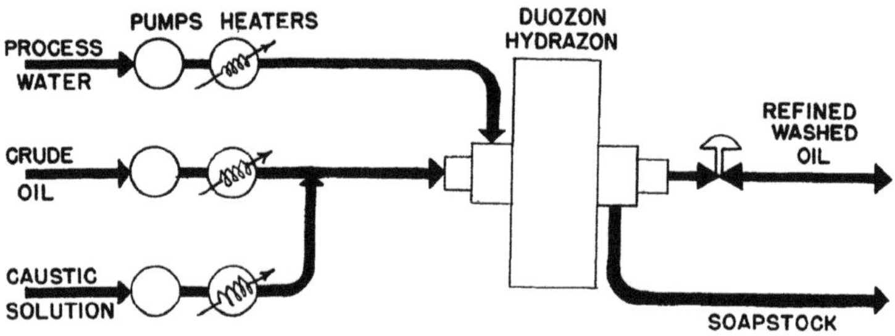

FIG. 6:4. ONE-STEP CONTINUOUS PROCESS FOR REFINING AND WASHING COCONUT OIL
(*Podielniak Div., Dresser Industries, Franklin Park, Ill.*)

Over-all losses, during refining, bleaching, and deodorizing, in the processing of crude oils with free fatty acids from 3 to 9%, are reported as 1.2 times the free fatty acids of the crude. The losses for the refining and water-washing, are reported as being 0.85 to 1 times the f.f.a. of the crude.

Typical losses in refining methods using kettles are 1.3 to 1.6 times the f.f.a.

Advantages of Single-Step Process

The advantages claimed for the Duozon-Hydrozon method of caustic-refining and water-washing of coconut oil are:

1. More efficient separation of oil and soap-stock.
2. Increased yields of high-quality oil.
3. Simplified operation—compact plant layout.

4. Versatility—the same machine can be used for degumming, caustic-refining, or a combination of caustic-refining and water-washing.
5. A completely hermetic system, with pressure discharge of streams of both refined oil and soap-stock.
6. Simplified cleaning, without dismantling the machine.
7. Low cost, virtually maintenance-free, continuous operation.

Swedish Oil-Refining Process

One of the features of an oil refining process, developed in Malmo, Sweden, by Margarin A. B. Zenith is that it produces a high yield of quality oil without bleaching.[2] Bleaching is side-stepped by preheating the oil under vacuum, then refining it, by gently bubbling oil in small droplets through a large column of weak lye.

This is a one-man operation in which the cost is said to be 30% below that of conventional methods.

The process is as follows: Crude oil is pumped from the supply tank to the top tray of the pretreatment vessel, at a pressure of 6 mm of mercury. The oil is deaerated, dried, heated to 175–180°F, then dropped into a second tray, where it is treated with 85% phosphoric acid. This treatment which lasts 30 minutes, eliminates impurities, such as gums, phospholipins, and other non-glyceride substances. The flow of oil is kept constant by circulating feed from the lower tray, used as a reservoir. It is neutralized in a vessel filled with 0.1–0.8*N* lye solution. A small quantity of surfactant is generally added, and the solution is heated to 175–180°F. The crude is introduced at the bottom of the vessel through perforated plates, which form uniform droplets.

As the droplets rise through the column of lye, the free fatty acids are neutralized, while the soap formed during the neutralization, remains dissolved in the lye. The fat coalesces into a layer on top of the lye and then flows to the next stage of the process. All traces of soap are removed from the neutralized oil, by treatment with citric acid, in a vessel at 6 mm mercury pressure.

At this point, diatomaceous earth is poured on the oil and it is recycled through the filter, after which the purified oil is ready for storage.

No bleaching is necessary in this process, which has been operating for several years on cottonseed, coconut, soybean, and hardened whale oil.

The Votator Division of Chemetron Corporation is negotiating to

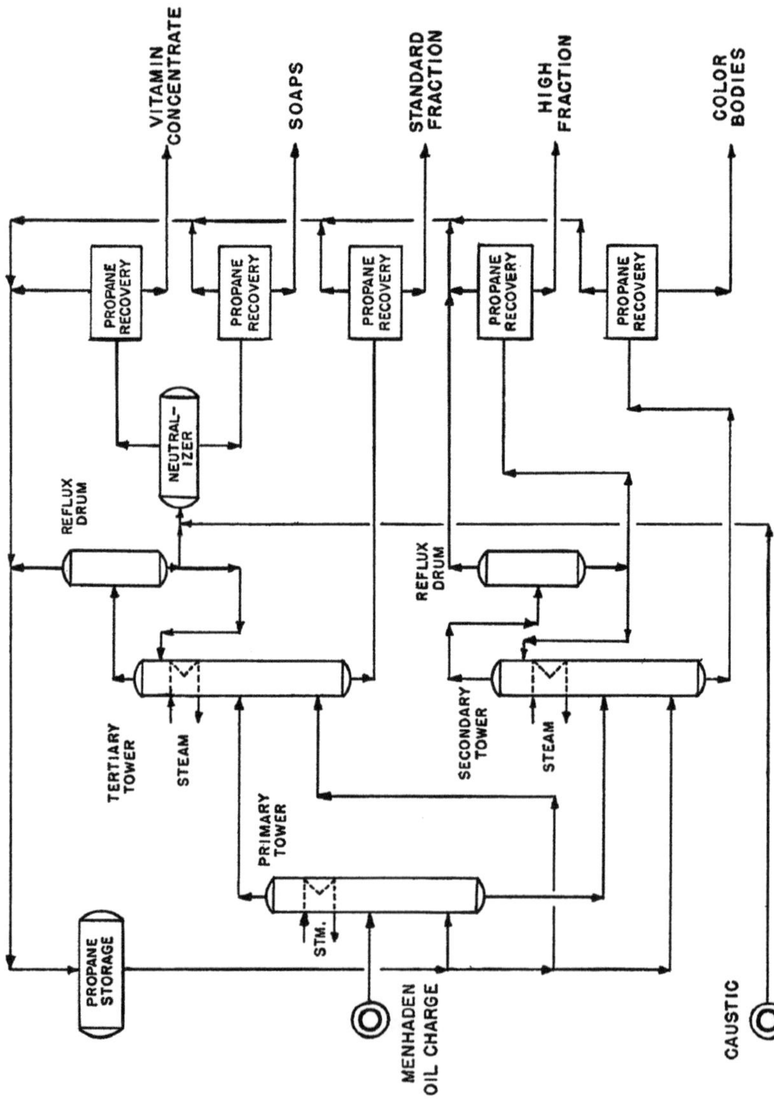

Fig. 6:5. Flow Diagram of Solexol Fractionation of Oil (*The M. W. Kellogg Co.*)

manufacture the equipment. The process will be licensed to domestic processors. All units necessary for a 5000 lb/hr Zenith refining plant can be installed in 240 sq ft, including processing vessels, pumps, equipment for agitation and filtration, plus controls.

The Solexol Process

A continuous method of refining oils was introduced by M. W. Kellogg Company in 1946 (Fig. 6:5). It is based on liquid-liquid extraction, using liquid propane as a selective solvent. At low temperatures, the charge glycerides are completely soluble in propane, but at higher temperatures the components of higher molecular weight are thrown out of solution into a separate layer. As the temperature is further raised, more and more of the heavier oils goes into the bottom layer. Finally, at the critical temperature of propane (206°F), practically every fraction has been precipitated.

TABLE 6:1

SOYBEAN OIL REFINED BY SOLEXOL PROCESS

				Further Refined		
	Crude	*Refined*	*Pigment Lecithin*	*Edible Oil*	*Paint Oil*	*Sterol*
Color-Gardner	12	3	Black	2	6	
Color-Lovibond		10Y-1R		10Y-1R	75Y-8R	10Y-1R
FFA as Oleic	0.4	0.2*	5.0	0.04	0.02	20.0
Unsaponifiable Oil, %	0.9	0.9	2.5	0.30	0.30	60.0
Sapon. No.	192	192	190	192	192	80.0
Iodine No.	135	135	100	125	162	115.0
% Acetone-Insol.			80			
Gardner Break	0.3					
Oil Yield, %	100.0	98.5	1.5	67.5	30	1.0

* Free fatty acids drop to less than 0.05% after deodorization.

Possible uses of the products:

Edible Oil: Shortenings, margarine, salad oils

Paint Oil: Paint, varnish, enamel, oil cloth, printing ink, alkyd resins

Pigment Lecithin: Emulsifying agent, chocolate coatings, pharmaceuticals, cosmetic preparations, oleomargarine, shortenings, leather dressings

Sterol Concentrate: It contains tocopherols and sterols useful in the synthesis of hormones and other pharmaceuticals

As applied to vegetable, animal, and marine oils, this process is said to have the following advantages over the older refining methods:

1. Highly unsaturated glycerides and fatty acids are not polymerized.
2. None of the vitamin content is destroyed by thermal decomposition.
3. Flavor and stability of the products are enhanced.

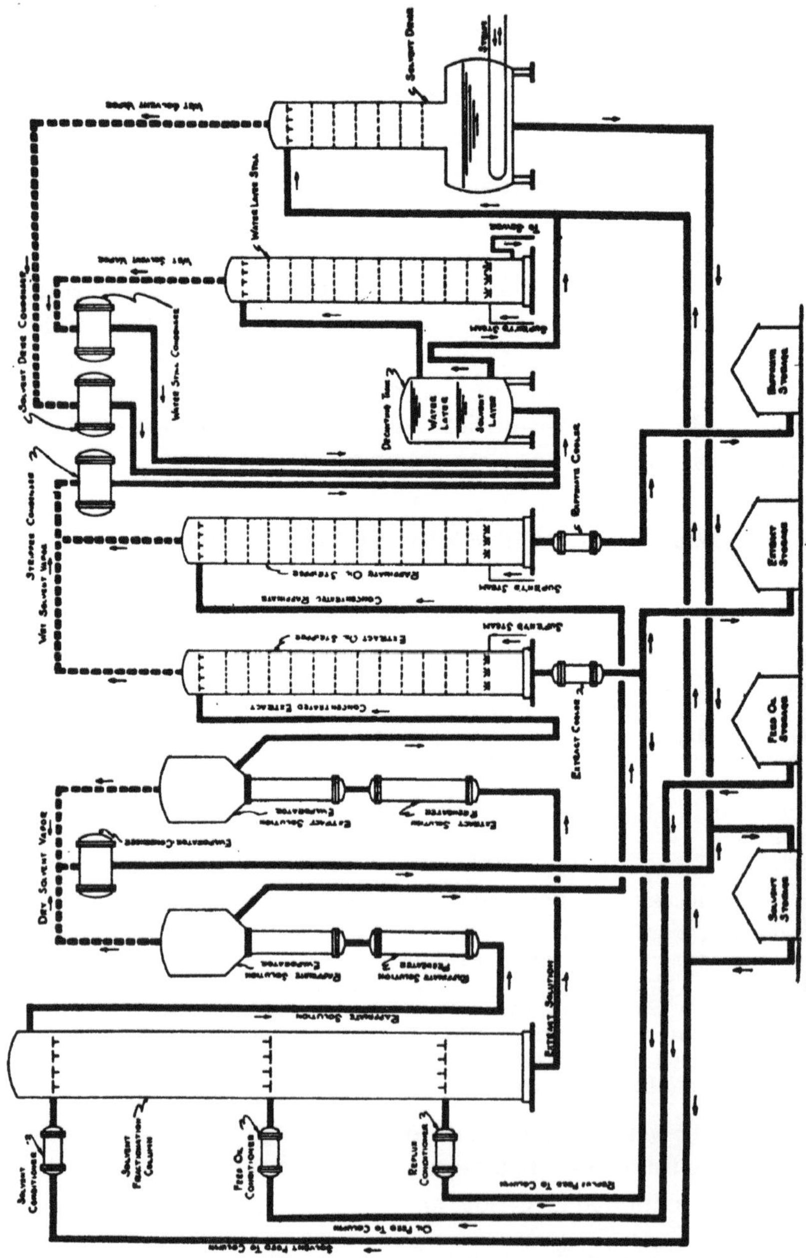

FIG. 6:6. FLOW DIAGRAM OF FURFURAL EXTRACTION PROCESS
(Reprinted with permission of *Journal of Industrial and Engineering Chemistry*)

4. All oil processed, regardless of source, remains in its original glyceride form.
5. Refining losses are avoided as there is no entrainment of glycerides by soap formation.
6. Vitamins A and D are recovered in their natural ester form.
7. Operating costs vary from a fraction of a cent to three cents a gallon.

Tables 6:1 and 6:2 show the physical and chemical properties of oils refined by this process.

TABLE 6:2

SARDINE BODY OIL REFINED BY THE SOLEXOL PROCESS

	Crude	Iodine No. Medium	Iodine No. High	Concentrated Vitamin Oil	Stearin	Colored Bodies	Fatty Acids
Vitamin A*	350			3500			
Vitamin D	100			1000			
F.F. Acids as Oleic	1	0.3	0.4	0.1	0.1	0.5	90
Iodine No. (Wijs)	185	160	240	120	50	250	180
Color (Gardner)	11	4	8	5	3	Black	8
Oil Yield, %	100	47	35	9	5	3	1

* International Units/g

Possible Uses of Products:
- Medium Iodine No. Oil: Paints, printing inks, extender for linseed oil
- High Iodine No. Oil: Varnishes, extender for China-wood oil
- Concentrated Vitamin Oil: Human consumption, fortifying of animal and poultry feed stocks
- Stearin: Shortenings and other edible products
- Colored Bodies: Source of compounds of high molecular weight
- Fatty Acids: Soaps, rubber, lubricants, resins

Furfural Process

This process for refining oils is based on the fact that when the oil and furfural are brought together in a fractionating column at a temperature lower than that of complete miscibility, two fractions are obtained.[5] This is accomplished in a packed counter-current column by passing the solvent downward through a rising column of oil. It has been found that the less saturated glycerides are preferentially soluble in the furfural-predominant phase,—the so-called *extract* phase—and the more saturated glycerides remain in the oil-predominant phase—the so-called *raffinate* phase. This process, however, is not limited to separation based on unsaturation within the glyceride molecule, but it operates also on differences in molecular weight, such as the length of the carbon chain, the fatty acids within the glyceride molecule, or on the differential solubility of the various minor ingredients in the oil.

For each part of oil six to fourteen parts of furfural is employed. The process has been adapted by the Pittsburgh Plate Glass Company to the fractionation of many vegetable and animal oils, and it has been operated commercially on linseed and soybean oils. Linseed oil has been fractionated to yield an extract resembling perilla oil; soybean oil, to yield an extract equivalent to linseed oil in drying and a raffinate which is an improved edible oil. The process is not limited to oils, but may also be used for fractionating fatty acids, for concentrating valuable minor ingredients of oils, and for separating compounds of different molecular weight.

In the concentration of vitamin A from liver oils by this process, a yield of 19%, containing 82,000 units to the gram, was obtained from dogfish liver oil which originally had a potency of 17,500 units.

Tall oil, which normally consists of about 50% of fatty acids, 42% of rosin, and 6.5% of unsaponifiable fat, is first esterified and then fractionated with a mixture of furfural and naphtha. By this process, two fractions are obtained. One contains only 1.3% of rosin acid and 2.9% of unsaponifiable fat, and the other consists of fatty acid esters. Figure 6:6 is a flow sheet of a plant handling one tank car of oil a day by this process.

Bleaching of Oils

The most important method of bleaching animal and vegetable oils is decolorization by means of adsorbents such as fuller's earth and activated charcoal.

The oil is transferred to a tank which may be similar in construction to the refining tank. Although it is not absolutely necessary to have a perforated coil, it may be useful for admitting steam while cleaning the tank or for stirring or even bleaching with air.

The oil is heated at 200–212°F to remove the last traces of moisture from the refining. As bleaching agents, 4% of fuller's earth and a smaller quantity of activated charcoal are added, and the oil is slowly agitated by means of a stirring device, for 15 to 30 minutes, to keep the bleach in suspension. The oil and bleach are then pumped through a plate filter press. Some refiners use broad paddles mounted on a horizontal axis, to prevent the bleach from settling.

Chemical Bleaching Methods

Edible palm oil which is bright orange-colored, can be bleached by a chemical method. It is heated to 190–200°F and air is blown through, by means of the open coil. In a few hours, the oil is almost completely decolorized. This process is employed also on such low-grades as Lagos and Niger palm oils. Sometimes chromic acid is used as the oxidizing agent. Sodium dichromate and sulfuric acid can be employed in a lead-lined, or other acid-proof, tank. Other oxidizing agents used are sodium hypochlorite, sodium chlorite (Textone*), hydrogen peroxide and other peroxides and persulfates.

When oxidizing bleaches are inoperative, or for oils that do not respond to oxidizing treatment, reducing agents may be used, such as hydrogen in the presence of a catalyst, sodium hydrosulfite, and sulfur dioxide.

Deodorizing

The principle of deodorizing oils depends on vacuum distillation of the odoriferous impurities with superheated steam. The deodorizer consists of a tank heated by a high-pressure, closed steam coil; or by a coil through which very hot oil from a special furnace is passed. Attached to the tank is a condenser and a vacuum pump.

The batch of oil, varying from 5000 to 30,000 lb, is heated to about 400°F and then the superheated steam is blown through it for three to six hours, depending on the type of oil and the vacuum. The odor-bearing impurities are carried off to the condenser. The vacuum is maintained until the oil is cool enough to be exposed to air without oxidizing. There is a loss of oil ranging from 0.12–0.35% with cottonseed oil, to 0.35–0.75% with corn oil and soybean oils. (see Fig. 6:7.) [6]

There are also some continuous-process deodorizers.

Winterizing

Oils can be winterized to free them of stearin. When cottonseed oil, or any oil containing a considerable amount of stearin, is chilled, e.g., by putting in the refrigerator, the stearin crystallizes, settles to the bottom of the container and cannot be poured. This is the principle of the winterizing process. The oil is chilled to 30°F and then filter-pressed to remove the crystallized stearin, which is used for the manufacture of shortenings. Oil that has not been chilled and pressed is sold as "prime summer yellow oil."

---

* Supplier: 86.

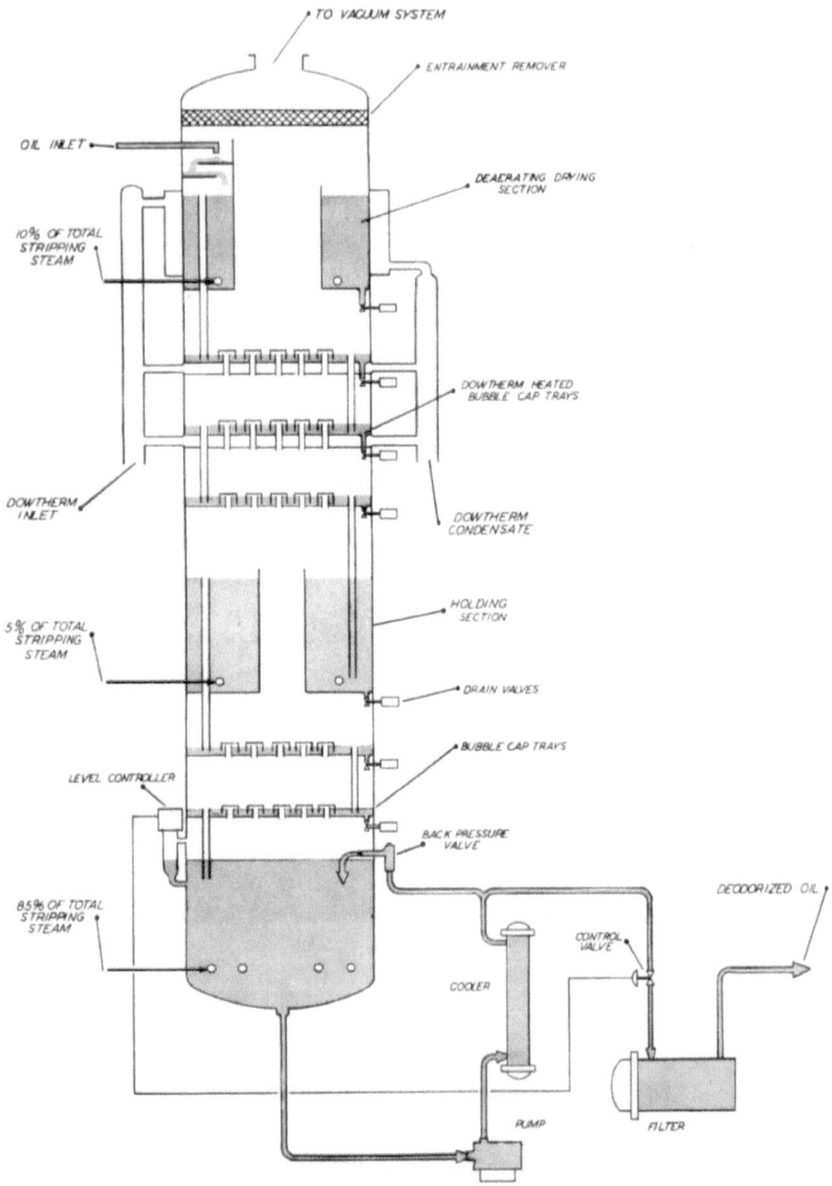

FIG. 6:7. CONTINUOUS DEODORIZING SYSTEM FOR OILS
(*Wurster & Sanger, Inc., Chicago, Ill.*)

Blown or Bodied Oils

When semi-drying, drying, or even fish oils are blown with air and heated, they become viscous. These blown oils and bodied oils are used in paints, varnishes, and lubricants.

The oils that consist chiefly of unsaturated fatty glycerides, when blown or heated at high temperatures (500–600°F), become oxidized. The higher the temperature and the longer the duration of heating, the more viscous the oil will be. Such oils, when used in paint, enamel, or printing ink, dry quickly.

In blowing oils for use in paints, one of the limiting factors is that as the oil becomes oxidized, its solubility in mineral spirits decreases. For special uses, the oil must meet tests for solubility as well as for viscosity or body.

Polymerization of Oils

Polymerization is a chemical change involving molecules of the same substance and resulting in the formation of a new compound whose molecular weight is a multiple of that of the original substance. The original compound is called a *monomer* and the final product a *polymer*.

Hilditch,[7] found that on distillation of a polymerized linseed oil in a molecular still, at very low pressure, the average molecular weight of the 30% fraction was 760. This portion consisted of oleic acid and the saturated acids that were in the original oil. The 70% portion, which remained in the still, showed an average molecular weight of 3500, i.e., it was polymerized.

The specific gravity of raw linseed oil rose from 0.9321 to 0.9741 upon being heated. The iodine number decreased from 169 in the raw oil, to 86 in the polymerized oil.

In China-wood oil or oiticica oil, which have three conjugated double bonds, gels are formed on ten minutes' heating. In the bodied oil, the conjugated unsaturation was found to be much lower than in the original oil. From this, it must be concluded that polymerization occurs at the conjugated unsaturated groups.

SPECIAL TREATMENTS OF OILS

Castor Oil

Dehydration of castor oil is a development of some years ago. This oil, which belongs to the nondrying class, is rearranged molecularly, so that it becomes a good drying oil.

Originally, the fatty acids (ricinoleic) were split out of the oil. It was then dehydrated and subsequently esterified with glycerol:according to the patents of J. Scheiber.[11] Later Hans Ufer [12] described a process in which castor oil is heated in the presence of a nonoxidizing acid and in the absence of any oxidizing agent.

Castor oil has a hydroxyl (OH) group and only one double bond. When dehydrated, the hydroxyl (OH) group and an adjacent hydrogen atom are removed in the form of water, thus opening up a second double bond, as shown in the following two possible structures:

$$\begin{array}{l} \quad\quad\quad\quad\quad\quad\quad\quad\quad\;\; \mathrm{H\;\;\;H\;\;\;H\;\;\;H\;\;\;\;\;H} \\ \mathrm{HOOC{-}(CH_2)_7{-}C{=}C{-}C{-}C{-}C{-}(CH_2)_4CH_3} + \text{dehydrating agent} \\ \quad\quad\quad\quad\quad\quad\quad\quad\quad\quad\quad\quad\;\; \mathrm{H\;\;\;OH\;\;H} \\ \quad\quad\quad\quad\quad\quad\quad\quad\quad\;\; 9\;\;\;10\;\;11\;\;12\;\;\;13 \end{array}$$

Ricinoleic Acid

$$\begin{array}{l} \quad\quad\quad\quad\quad\quad\quad\quad\quad\quad\quad\quad\;\; \mathrm{H\;\;\;H\;\;\;H\;\;\;H\;\;\;H_2} \\ \rightarrow \mathrm{HOOC{-}(CH_2)_7{-}C{=}C{-}C{=}C{-}C{-}(CH_2)_4CH_3} \\ \quad\quad\quad\quad\quad\quad\quad\quad\quad\quad\quad\quad\;\; 9\;\;\;10\;\;11\;\;12\;\;13 \end{array}$$

Conjugated Acid

or

$$\begin{array}{l} \quad\quad\quad\quad\quad\quad\quad\quad\quad\; \mathrm{H\;\;\;H\;\;\;H_2\;\;H\;\;\;H} \\ \mathrm{HOOC{-}(CH_2)_7{-}C{=}C{-}C{=}C{-}C{-}(CH_2)_4CH_3} \\ \quad\quad\quad\quad\quad\quad\quad\quad\quad\; 9\;\;\;10\;\;11\;\;12\;\;13 \end{array}$$

Linoleic Acid

In explanation of this change, Mattiello [8] says that from a chemical point of view, it appears that dehydration puts a double bond in the 11–12 position. This would theoretically yield 100% of the conjugated isomers as claimed in the Scheiber patents. Analysis does not support this claim. It is equally possible for the dehydration to occur in the 11–12 position and in the 12–13 position. On this basis, the dehydrated castor fatty acids would theoretically be composed of 42.5% each of the conjugated and nonconjugated isomers, and 15% of other acids. Therefore, one would assume that the properties of dehydrated castor oil are between those of linseed and China-wood oil, which is actually the case.

Priest and von Mikusch,[9] found that the dehydrated castor oil of commerce usually contains 17 to 26% of the conjugated and 59 to 64% of the nonconjugated acid.

Alexander Schwarcman's patent [10] describes a process in which castor oil is heated at 170–260°C with a suspended catalyst, until there is a large increase in the iodine number. The catalyst is then removed and the heating continued at 285–315°C to stabilize the oil and develop drying properties. The catalyst consist of a mixture of fuller's earth, which has no polymerizing or hydroxylating effect on castor oil, with minor amounts of sulfuric or acetic acids or sodium acid sulfate. Other processes use phosphoric acid.

Although the drying properties of dehydrated castor oil may lie between those of linseed and China-wood oil, its color is superior to that of both. It is sold under such trade names as Scheiber Oil, Isoline, Dienol, Castung, and it is also known as octadecadienoic glyceride. The iodine number of dehydrated castor oil is about 141 (Wijs).

Linseed Oil

In an older process of refining (see also Chapter 3), the oil is treated with 1% of its weight of sulfuric acid, in a lead-lined tank provided with an agitator. The acid is sprayed into the oil while it is being stirred, and the oil turns a muddy green. As the next step, 1% of water (based on weight of oil) is sprayed in, to arrest the action of the acid. The coagulated matter is allowed to settle, the oil is washed twice with water, and is agitated with steam. The batch is then pumped into a settling tank. Oil refined by this process is used in wet pulping of white lead, for grinding pigments and in the manufacture of thick lithographic inks. This process has been almost entirely superseded by the alkali-refining method, which gives better results.

Oleomargarine

The process used at present in the manufacture of oleomargarine (see also Chapter 2), consists in churning vegetable oils and or fats sometimes with skimmed milk which has been ripened with lactic acid bacteria. This produces an artificial buttermilk, which imparts the flavor and odor of butter to the fats.

Definitions and standards set up under the Federal Food, Drug and Cosmetic Act, in the U. S. Department of Health, Education and Welfare, Part 45, stated briefly, are, that oleomargarine may be prepared from either animal or vegetable fats or oils, or a combination of both, some of which may be hydrogenated; cream, milk (cow's) skim milk or any combination of nonfat dry milk and water, not less than 10% solids, or in the case of predominantly vegetable margarine, a milk made of 10% of ground soybean and 90% of water; the mixture is congealed either with or without contact with water and the congealed mixture may be worked.

Provitamin A shall be deemed to be artificial coloring.

Sodium benzoate or benzoic acid, or a combination of both, not in excess of 0.10% by weight of the finished product, may be used as preservative.

Vitamin A, with or without any accompanying vitamin D, and with or without vitamin D concentrate, may be used in such quantity that the finished oleomargarine shall contain not less than 15,000 USP units to the pound. The vitamin potency may be furnished: By fish liver oil, by concentrates of vitamin A or its fatty acid esters from animal sources; by synthetic vitamin A or its fatty acid esters; by mixtures of synthetic vitamin A or its fatty acid esters with harmless substances formed during the synthesis of vitamin A, if the vitamin or its fatty acid esters constitutes not less than 50% of the mixture; by provitamin A or by any combination of two or more of these. (For the purposes of this subdivision, the term *fatty acid* may include acetic.)

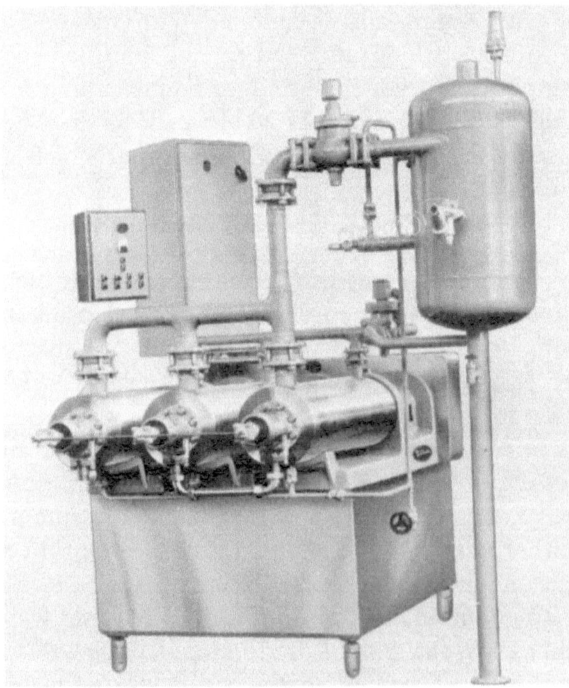

FIG. 6:8. VOTATOR FOR MANUFACTURE OF OLEOMARGARINE
(*Votator Div., Chemetron Corp., Louisville, Ky*)

The artificial flavoring *diacetyl*, may be added as such or as starter distillate, or produced during the preparation of the product as a result of adding citric acid or harmless citrates.

Lecithin, not exceeding 0.5%; mono- and diglycerides not exceeding 0.5%.

Butter, salt, citric acid may be incorporated in the fat or oil ingredient used. Isopropyl citrates are not to exceed 0.02%; stearyl citrate not over 0.15 per cent. The finished oleomargarine contains not less than 80% of fat. The ingredients must be marked on the label.

In the early days of manufacturing oleomargarine the whole charge of emulsified oils was run through a chute of cold water, into boxes with slotted sides so that the excess water ran off readily.

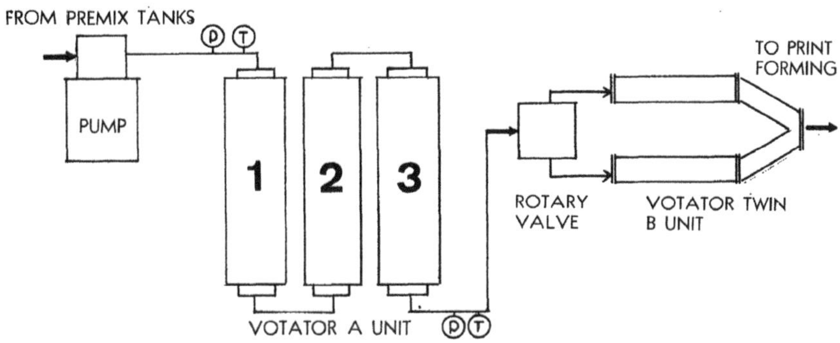

FIG. 6:9. DETAILS OF VOTATOR
(*Votator Div., Chemetron Corp., Louisville, Ky*)

At present, however, continuous emulsification [13,14] is effected by pumping the accurately proportioned ingredients into a blender and from there through a Votator (Fig. 6:8) which consists of one or three closed cylinders (*A* unit), which are six inches in diameter and 48 inches long. The fat enters around 100°F, and in 16 seconds it is cooled to about 50°F. The mixture at this stage is quite fluid, because it is in a supercooled condition. From there it is forced into the *B* unit (see Fig. 6:9). The *B* unit is 7 inches in diameter and is made up of flanged sections, 18 inches long to permit the length to be varied in accordance with the physical characteristics of the fat that is being solidified. From here, it goes to the print-forming machine. The product is not exposed to the atmosphere until it reaches the print-forming equipment.

It is worth noting that ever since the scare about adverse physiological effects from using saturated fats for edible purposes, accent has been put on the polyunsaturated portion of vegetable oils. Some manufacturers of margarine, following this line of thought, are now advertising that their margarine is "100% Dairy-Free," to indicate complete exclusion of butter-fat. Additionally, butter-free margarine attracts the "kosher trade," because the margarine may then be used both in dairy and in meat foods. (See section on *Polyunsaturates* in Chapter 3.)

Hydrogenation of Oils

A fluid oil such as cottonseed oil, can be hardened to a solid, white fat, similar in appearance to tallow, by means of hydrogenation.

The hydrogenation process was developed after Sabatier and Senderens, in 1896, published their work on the effectiveness of nickel and other metals, as catalysts for hydrogenation.

W. Norman, in 1902, described a process for hydrogenating oil or fatty acids in either the liquid or vapor phase, in the presence of a nickel catalyst, supported on pumice.[15]

At the same time Leprince and Siveke in Germany were also working on the hydrogenation of oil.[16]

Bedford and Williams were granted a British patent in 1907, covering a process in which the oil was sprayed into an atmosphere of hydrogen in the presence of finely divided nickel, which acted as the catalyst.

In the United States Kayser obtained patents in 1911, which describe methods for agitating the oil and a catalyst, under pressure of hydrogen, and the use of kieselguhr as a carrier for the catalyst. These patents were assigned to Procter & Gamble Company, and are employed in the manufacture of Crisco.[17]

At present, not only oils are hydrogenated on a vast scale, but also resins, terpenes, alkaloids, vitamins, mineral oil, and coal.

Pure hydrogen may be obtained by the electrolysis of water or by passing steam over red hot iron sponge. In some special processes, hydrogen is produced by reacting metallic sodium with alcohol.

Although platinum, palladium, copper, and iron may be used as catalysts, nickel has been found the most efficient and economical. The preparation of finely divided nickel, deposited on the proper carrier, is very important. To increase the surface of the catalyst, the nickel is precipitated on pumice, sawdust, kieselguhr, or charcoal.

Preparation of Catalyst

Nickel oxide or hydroxide can be prepared by the ignition of nickel nitrate; or nickel hydroxide can be precipitated directly on the carrier to be used.

The oxide is carefully reduced to metallic nickel at a temperature of 325°C. This is accomplished in a rotary drum into which the hydrogen is pumped under pressure. The finished catalyst is then immediately dumped into the oil to be treated, so that the catalyst is not poisoned by contact with atmospheric oxygen.

By another process, known as *wet-reducing*, the catalyst is prepared by treating nickel hydroxide or carbonate with formic acid to convert it to the formate. The latter is easily reduced to metallic nickel by heating at 150–190°C, while suspended in coconut oil or another saturated oil.

Murray Raney developed a process for the preparation of nickel catalyst, in which an alloy of nickel and aluminum is reacted with aqueous sodium hydroxide. The excess soda and sodium aluminate are washed away and the nickel is dried in oil. This process is covered by a patent,[18] as are improvements of Raney nickel, on the preparation of a lump-type catalyst applicable to the hydrogenation of rosin and benzene;[19] and the recovery of spent catalyst.[20]

Process of Hydrogenation

The hydrogenator is a closed tank, equipped with an agitator or pump to circulate the oil and catalyst. The hydrogen is pumped under pressure, while the oil is maintained at 175–190°C.

Cottonseed and whale oil require 1000 cubic feet of hydrogen for a ton of oil.

As the hydrogenation proceeds, the iodine number of the oil is reduced, while the titer and melting point become higher than they were originally. The odor of the finished product is usually improved.

Hydrogenated oil may be produced in various consistencies ranging from that of lard, to such hardness as is suitable for pulverizing. Hydrogenated oils are used in edible products, as substitutes for waxes, in compounding rubber and phonograph records, and for many other purposes.

When a sample indicates that the oil has been hydrogenated to the proper hardness, it is filtered to remove catalyst and carrier. It is now ready for use.

Whale oil is fluid and has a fishy odor even when fresh. It may be hydrogenated to a hard, white fat, which has lost most of its odor during the process. Much of this oil goes into the manufacture of soap as a substitute for tallow. In recent years, however, tallow is so cheap that no substitute is needed. About half of our present production of tallow is exported. A comparison of the treated and untreated whale oil, shows the following constants:

Type	*Titer*	*Iodine No.*	*Color*
Untreated Oil	23°C	90–146	Yellow-brown
Treated Oil	42–80°C	36–6	White

Sulfonation (Sulfation) of Oils

The National Bureau of Standards, after a number of conferences with manufacturers and users, promulgated, on January 26, 1933, *Commercial Standard C-S-43-32*, for the guidance of producers, distributors, and users, in order to avoid confusion resulting from diversity of opinion on the grading of sulfonated oils. It defines this commodity as follows:

> The term sulfonated oil, shall designate the product of the interaction of a saponifiable oil or fat or their fatty acids and sulfuric acid or similar sulfonating agent, the reaction to take place under such conditions that some or all of the oil is converted into a sulfate. With the exception of water and alkali, it shall contain no other admixture. A sulfonated oil of any particular designation shall contain no other oil or fat except the kind designated.

The history of sulfonated oil dates from Fremy (1831) and Runge (1834) who described oils that had been mixed with sulfuric acid. A British patent was granted to Mercer, in 1847, for sulfonating olive oil, which was to be used in dyeing madder Turkey reds. The name "Turkey Red Oil" has since been applied to sulfonated castor oil, which was first used in 1870.[21]

These oils should contain the minimum of free SO_3. If the oil is not properly washed and settled after sulfonation, considerable amounts of ammonium or sodium sulfate will be present. The salts are determined as uncombined SO_3.

The reaction between any oil and sulfuric acid takes place in several ways, depending on the temperature, the intimacy with which the materials are brought into contact, and the time. The major reaction results in a *sulfated* rather than a sulfonated product. With ordinary oils, sulfation occurs at the double bonds of the fatty acids, resulting in triolein hydrogen sulfate. Further changes yield hydroxy compounds and lactones.

Sulfonation Process

The sulfonation process consists of charging the oil into the sulfonator, which may be a steel-jacketed, lead-lined tank with an agitator, or a Monel-lined tank—which was found to be superior. The tank may be single-shell type, equipped with a noncorrodible cooling coil. The oil is agitated and cooled to room temperature or even lower. Concentrated sulfuric acid, 25% on the weight of the oil, is added slowly. The temperature of the mixture is maintained below 100°F. Experience has shown that oils containing about 4% of combined SO_3 can be used for blending with either vegetable or mineral oil, whereas those that are more highly sulfonated, do not blend.

If a blending oil is to be produced, the entire amount of acid is added to the oil within 15–30 minutes, but for higher sulfonation, the mixing will require a longer period.

Sometimes, after the reaction has taken place, the mixture is allowed to stand for several hours or overnight. It is then run into the wash tank, which may be similar to the sulfonator tank. Here it is mixed with an equal volume of water and allowed to stand until the batch has separated into an oily top layer and a bottom layer consisting of an aqueous acid solution, which is drawn off. One or two washes with brine (1.5 lb salt to a gallon) or Glauber salt solution follow. Each time, the brine is thoroughly mixed with the oil and allowed to settle. The object is to wash out as much of the uncombined sulfuric acid as possible.

At this point, manufacturers differ as to the mode of procedure, some neutralize with aqua ammonia, while others use 24 Bé solution of caustic soda, which is added with constant stirring, until a clear, transparent oil results. The *p*H of this oil should be between 5 and 7. The finished oil should be allowed to stand several days before drawing off, in order to allow any water, containing sulfates, to settle out. If this water settles in the shipping drums, the sulfates occasionally crystallize out after the oil has been run off, especially when the oil is subjected to cold weather conditions for a considerable period.

FAT-SPLITTING

The decomposition of fats into their fatty acids and glycerol (Fig. 6:10), called fat-splitting, has been carried on for many years. Although the reaction between a fat and an alkali may be considered a fat-splitting process, it yields soap and glycerol. Other types of fat-splitting (hydrolysis) yield more glycerol, and the fatty acids can be used for purposes other than soap.

FIG. 6:10. PLANT FOR MANUFACTURE OF FATTY ACIDS
(*Emery Industries, Cincinnati, Ohio*)

At first, the main purpose of fat-splitting was to provide fatty acids for the candle industry, but as technology of fats advanced, many uses for fatty acids as well as for glycerol were found. At present, fatty acids are used for candles, soap, for the production of alcohols, esters, amines, amides, and a host of derivatives. Mixed fatty acids are fractionated in molecular stills and other types of vacuum stills, to yield fatty acids with definite physical and chemical properties.

By whatever process the fat may be split, the chemical reaction is always the same:*

$$
\begin{array}{lcccccc}
C_{17}H_{35}COO\,CH_2 & & H_2O & & C_{17}H_{35}COOH & & CH_2OH \\
\quad\quad\quad\quad\;\; | & & & & & & | \\
C_{17}H_{35}COO\,CH & + & H_2O & = & C_{17}H_{35}COOH & + & CHOH \\
\quad\quad\quad\quad\;\; | & & & & & & | \\
C_{17}H_{35}COO\,CH_2 & & H_2O & & C_{17}H_{35}COOH & & CH_2OH \\
\text{Glyceryl Stearate} & & \text{Water} & & \text{Stearic Acid} & & \text{Glycerol}
\end{array}
$$

There are several processes by which fatty acids can be isolated from the glycerol.[22] According to Wurster & Sanger, Inc., chemical engineers of Chicago, who have been installing fatty-acid plants for many years, present demands are for maximum split (97–100%), minimum unsaponifiables, best color, and greatest color-stability.[23] Autoclaves for 150 to 600 psig have been displacing the old-time Twitchell plants.

In plants where one type of fat predominates, continuous splitting is recommended. Where various fats must be split, batch- or semi-continuous autoclaves work out best, because of their flexibility. Continuous splitters require 12–18 hours or more to change from one feed stock to another. No time is lost with batch- or semi-continuous autoclaves.

Continuous Autoclave Process

Decomposition in autoclave under high pressure, naturally entails high cost of installation, but it avoids the use of a catalyst which must later be freed from the product. A description of the semi-continuous autoclave process at 600 psig may be in order.

The advantages of this process are that 98–100% splits and increased concentration (up to 18% of glycerol sweet-waters) are possible, so that the operation is definitely economical also. The yield of glycerol is higher. Furthermore, if the fatty acids produced are to be reacted directly, no neutral stock is present to delay or interfere with the reaction. If the fatty acids are distilled, the yields are higher because of the decreased amount of still bottoms.

* Most oils or fats yield a mixture of fatty acids and glycerol.

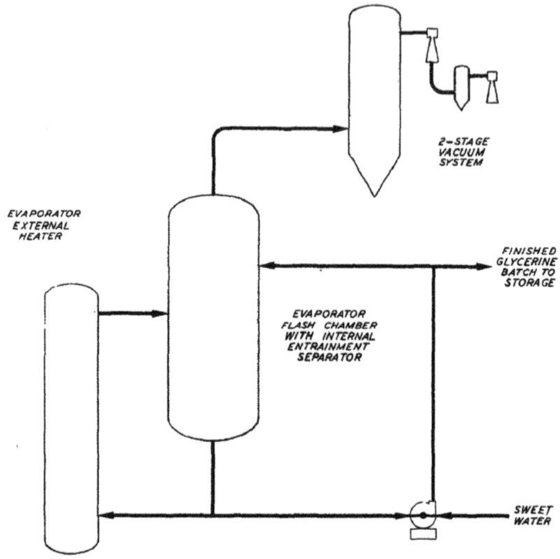

BATCH PROCESS, EVAPORATION OF SWEET WATER

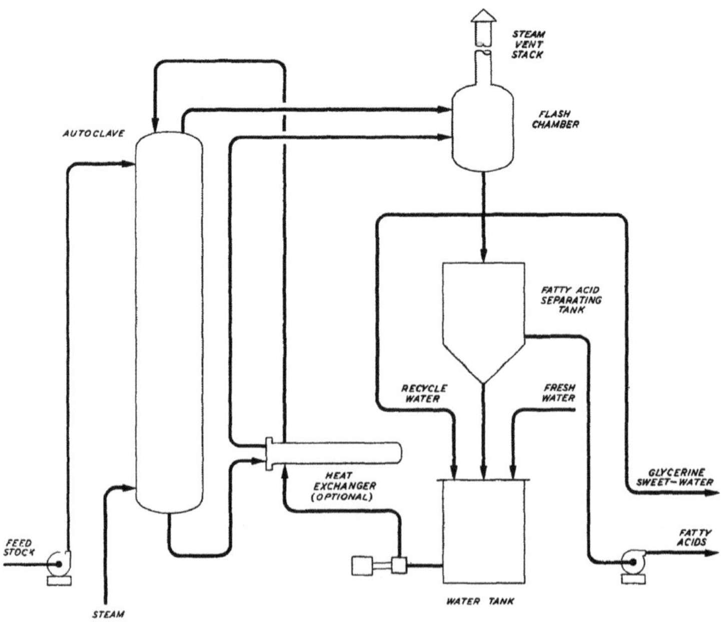

FIG. 6:11. SEMI-CONTINUOUS PROCESS FOR FAT-SPLITTING
(*Wurster & Sanger, Inc., Chicago, Ill.*)

In operation, fat is charged to the autoclave and steam injection started. Most of the steam will condense until operating pressure is reached. At that point, dilute sweet-water from a previous run is pumped into the autoclave under pressure and at a metered rate. At the same time, sweet-water is discharged automatically from the autoclave at the same rate at which water is fed to it. This is accomplished by a form of liquid level control which measures the head of liquid in the autoclave and maintains it constant. Special devices are built into the autoclave to effect a separation of fatty acids and sweet-water at the bottom. Thus, there is set up a system wherein the fat charge remains in the autoclave until it is completely split while splitting water flows through continuously. Such an operation is termed *Semi-Continuous;* see Figure 6:11.

When dilute sweet-water from a previous run is consumed, fresh water is fed in its place until the content of glycerol in the sweet-water discharged drops to 2–3 per cent. At this concentration, the equilibrium of the reaction is thrown completely in the direction of splitting and the resulting split is complete. The feeding of water is stopped and the autoclave is discharged. It is then ready for another cycle of operation.

The economy of this operation is in the subsequent step of evaporating the glycerol. The usual glycerol content of autoclave sweet-water is 10–12 per cent. By increasing this to 18%, the evaporation load—i.e., the steam consumed in evaporating a fixed amount of glycerol—is sharply reduced. Add to this the advantages of a complete splitting of the fat and the merits of the process become worth considering, particularly for production in the range of 10 tons or more a day.

Despite the advantages of the continuous autoclave method, the Twitchell Process for fat splitting will probably always be used because of its simplicity and initial low cost. It at least enables a small processor to produce fatty acids, but its over-all economics are less favorable today as compared with any of the autoclave methods. It is well to review it here.

Twitchell Process

The apparatus used for the Twitchell process consists of a lead-lined tank provided with a perforated lead steam coil. The fat is heated to 212°F with open steam. One percent of concentrated sulfuric acid (60°Bé) is added to a pure fat, such as tallow; or 3–4%

to garbage grease. This preliminary boiling which is continued for an hour, is a purifying process, during which organic matter is coagulated and precipitated. The contents of the tank are allowed to settle overnight and the fatty layer is drawn off into the splitting tank. The spent acid is drawn off into another tank, from which any residual fat may be skimmed.

The splitting tank is constructed of 3-inch cypress staves, with a close-fitting cover and a capacity of twice the volume of the fat to be split. It is equipped with a steam coil at the bottom and another steam pipe which provides an atmosphere of steam over the surface of the fat, so as to keep the surface free of atmospheric oxygen, which darkens the fat. A tank having the splitting capacity of 10,000 lb should be 7 feet 6 inches deep and 7 feet 10 inches in diameter.

To the purified fat in the splitting tank, the following mixture is added.

Water	40%
Saponifier	0.6–1
Sulfuric Acid (60°Bé)	0.7–1

The Twitchell Contact Saponifier consists of sulfo-aromatic stearic or oleic acid. The aromatic portion, once naphthalene, has been displaced by cymene, anthracene, and other aromatics.

These are boiled for 18 hours with open steam. A test shows that at this point the fat contains 88% of free fatty acids. The glycerol (sweet-water) is separated, but an atmosphere of steam is maintained over the surface of the fat.

Fifteen percent of fresh water, with 0.2% of 60°Bé acid, is added and the boiling is resumed for 12–15 hours, during which time the decomposition will proceed to 95% or higher.

The sweet-water is withdrawn and the remaining traces of sulfuric acid are neutralized with soda ash, 0.5 lb/1000 lb acid present.

The fatty acids produced from pure fats, such as tallow or vegetable oils, are pure and of good color. They may be used directly in the soap kettle or for other purposes. Garbage grease and cottonseed foots yield fatty acids that are almost black or, at best a very dark green. Such acids must be distilled to make them suitable for soapmaking or other purposes.

Distillation and Pressing

Distillation of fatty acids was originally carried out in retorts, heated by direct fire. Superheated steam was blown through the stock in the retort to effect distillation, with little decomposition. This yielded a mixed fatty acid, yellow in color, from which three fractions were separated. The first was light in color, the second was darker, and the remainder was a dark tarry substance which was used for pitch. The fatty acids were used in making soap, or chilled and pressed for stearic and oleic acids (red oil).

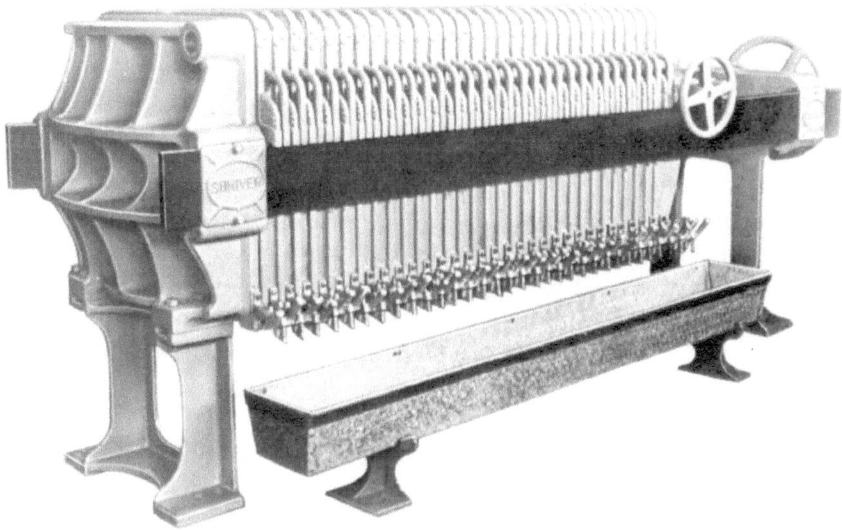

FIG. 6:12. FILTER PRESS FOR FATTY ACIDS
(*T. Shriver & Co., Inc., Harrison, N.J.*)

If the stock is to be pressed, it is melted and run into specially constructed aluminum trays which hold about ten pounds of fatty acids. The trays are set on shelves and, beginning with the top shelf, each tray is filled, allowing the overflow to run by means of a spout, into the tray below, until all are filled. The fat is then chilled to 40°F. As each tray is emptied, the slab of fatty acids is wrapped in a camel's hair press cloth or burlap, and set into a hydraulic press. The products are single-pressed stearic acid and red oil.

The stearic acid is melted and recast as before and then it is given a second pressing in a steam-heated press at 100°F (Fig. 6:12). This yields double-pressed stearic acid and more red oil. The process is repeated once more to produce triple-pressed stearic acid.

The characteristics of the three grades of stearic acid are as follows:

	Melting Point °F	*Iodine Number*	*Titer °C*
Single-pressed	125–126	15	53
Double-pressed	128–129	8	54
Triple-pressed	130–131	3	55

The color varies from straw yellow for the single-pressed to white for the triple-pressed grade. The odor is also improved by the repeated pressings, from a characteristically rancid odor in the single-pressed to practically nil in the triple-pressed grade of stearic acid.

Commercial stearic acid actually consists of about equal parts of palmitic and stearic acids, with minor proportions of oleic and other unsaturated acids.

The red oil of commerce has a ruby color by transmitted and a dark-brown fluorescence by reflected light.

It is to be noted that fatty acids produced by hydrolysis of tallow need not be distilled. Tallow yields a stearic acid of good color, and oleic acid, which is known as *Saponification Red Oil.*

Continuous Distillation

There are several processes that use continuous distillation.[24] In the simple distillation of fatty acids, improved results with respect to quality of product, uniformity, and yields are obtained with reduced requirements for plant operation and space because of the following factors:

1. Continuous operation under automatic instrument control.
2. Deaerating the feed stock before heating.
3. Vacuum-drying the crude fatty acids at low temperatures.
4. Indirectly heating the fatty acids to distillation temperatures by means of condensing Dowtherm vapors within the bubble trays of a stripping still. With this unique heating system the fatty acids are not exposed to high temperatures until they are under high vacuum in the still itself, thereby avoiding the preheating of feed stocks far above distillation temperatures before they are under high vacuum.
5. Low hold-up and minimum transfer of heat to temperature of surface in the stripping still.
6. Removal of entrained matter from vapor stream of distillate before condensing the product.

7. Efficient system for condensing product and cooling the product before discharging it to the atmosphere.
8. Wet scrubbing of the stream of vapor from exhaust after condensing the product.
9. Residues that are easily handled for the recovery of free and combined fatty acids.
10. Freedom from metallic contamination, because of all stainless steel construction wherever there is contact with the fatty acids.

Figure 6:13 shows the general arrangement of equipment of such an installation by Wurster & Sanger.

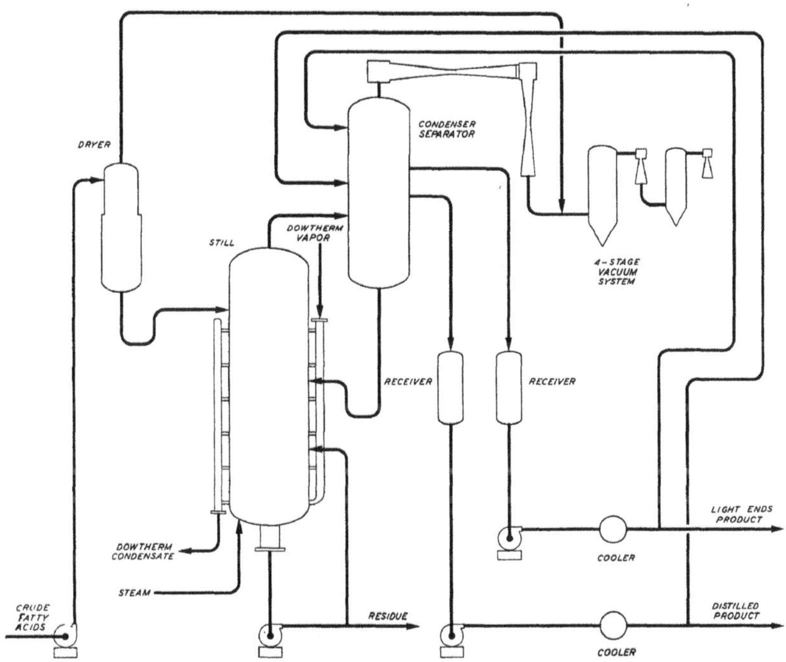

FIG. 6:13. CONTINUOUS DISTILLATION OF FATTY ACIDS
(*Wurster & Sanger, Inc., Chicago, Ill.*)

Fractional Distillation

As a result of fractional distillation, many previously unknown fatty compounds have been synthesized. Pure fatty acids with definite characteristics are produced commercially. By fractional distillation and by other means of separating the fatty acids derived from semi-drying and from marine oils, it has become possible to separate the drying from the non-drying fractions. By subsequent re-esterification of these fractions with glycerol and pentaerythritol, valuable drying oils can be produced for the manufacture of paints and varnishes. The production of many derivatives of fatty acids became possible by these developments.

FIG. 6:14. UNIT FOR DISTILLATION OF FATTY ACIDS
(*Armour & Co., at McCook, Illinois*)

Armour & Company holds patents covering processes of heating the fatty acids which are pumped into the vaporizing chamber, to which superheated steam is admitted (Fig. 6:14).[25] The chamber communicates with a fractionating column, where distillation takes place. The unsplit fatty acids are recycled from the bottom of the column to the vaporizer. Fatty acids of definite composition are tapped off at several places in the column and condensed. The entire apparatus operates under vacuum produced by steam ejectors and barometric condensers.

Table 6:3 shows the purity of fractionated fatty acids, as compared with pure stearic and pure lauric acid.

TABLE 6:3

COMPARISON OF FRACTIONATED AND PURE FATTY ACIDS

	Neofat No. 1-65[a]	*Stearic Acid*	*Neofat No. 11*[b]	*Lauric Acid*
Mean Molecular Weight	282.5	284.28	203.0	200.19
Melting Point, °C	67.0	69.9	37.8	43.6
Iodine No. (Wijs)	3.0	none	1.0	none
Neutralization Value	198.0	197.0	276.0	280.0
Unsaponifiable Matter, %	0.5	none	0.4	none
Color	white	white	white	white
Odor	tracc	none	trace	none

a Armour & Co., Chicago; composition: stearic, 90%, palmitic, 6%, oleic, 4%
b Armour & Co., Chicago; composition: lauric, 90%, capric, trace, myristic, 9%, unsaturated, 1%.

Emersol Process

The Emersol process separates solid from liquid fatty acids by means of a common solvent, from which the solid fatty acids are crystallized and filtered out of solution.

The mixture of distilled fatty acids is dissolved in methanol of the proper concentration, along with a small amount of crystal promoter, which is added by means of a small proportioning pump.

The solution is chilled in a multi-tubular, agitated crystallizer, which results in a slurry containing stearic acid crystals in suspension.

The crystalizer and cold room are refrigerated with a two-stage ammonia machine driven by a steam turbine. The steam from the exhaust is used for operating the stills elsewhere in the plant.

Leaving the crystallizer, the slurry flows to the feed pan of the totally enclosed rotary filter, installed in a refrigerated and insulated cold room. The design of the filter permits continuous operation with the minimum of maintenance. The filter cake is removed in strings. A washing with hot solvent maintains the filter cloth in proper condition.

The filter cake is transferred to a melter, from which it is pumped to a stainless-steel still, where the solvent is removed and recovered for reuse.

The filtrate containing the oleic acid and other liquid acids is drawn off from the filter and pumped through a multi-tubular economizer, which cools the alcohol entering the process and thus recovers

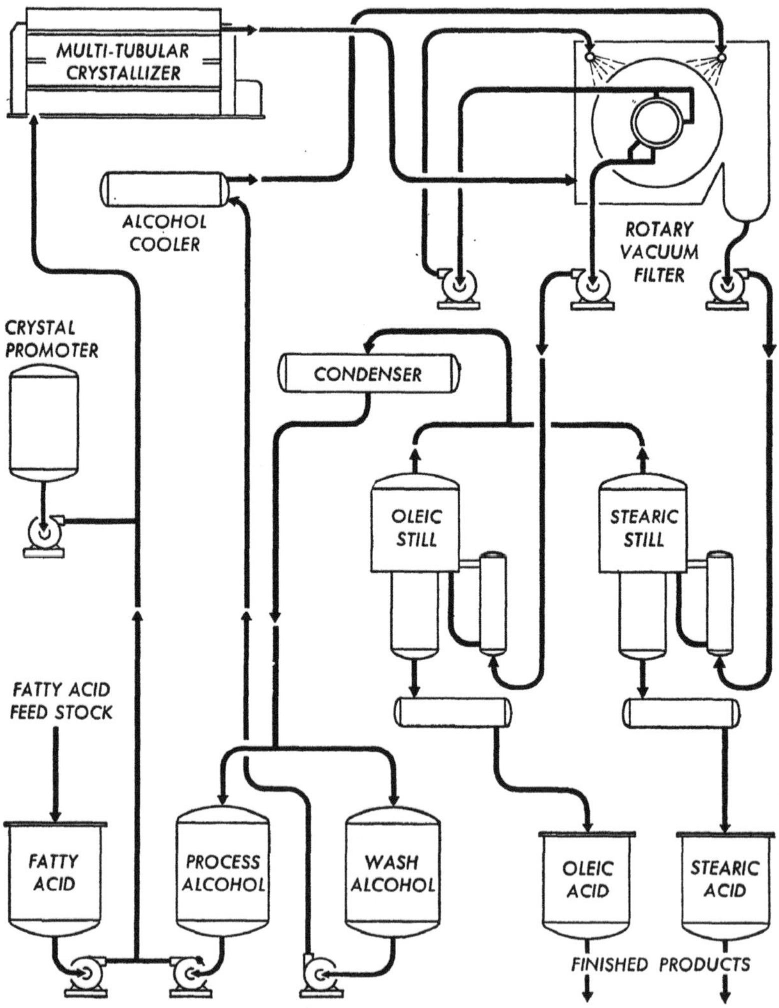

FIG. 6:15. EMERSOL PROCESS FOR SEPARATING FATTY ACIDS
(*Emery Industries, Cincinnati, Ohio*)

almost 75% of the total refrigeration. The solvent-acid mixture is pumped to the oleic acid still, which is made of stainless steel and designed to ensure complete removal of the solvent, under conditions that prevent esterification, overheating, and discoloration of the stock.

Everything is controlled automatically from a central panel. A plant with capacity of 30-tons a day (2500 lb of feed stock an hour) operates 24 hours a day, for two weeks, without requiring a shutdown. The plant is housed in a two-story building 50 by 60 feet. Both high- and low-grade fatty acids may be treated.

Figure 6:15 shows a simplified flow sheet of the Emersol process.

Hydrogenation of Fatty Acids

The De Nora process of high-pressure hydrogenation of fatty alcohols (see Chapter 4) has been used successfully by Procter & Gamble in California.

This process can be carried out as a coniinuous or semi-continuous process, depending on the capacity of the plant. The semi-continuous process is usually employed for quantities less than ten tons a day.

Figure 6:16 shows the flow sheet of the process. It gives high yields when the rate of circulating hydrogen is kept high, and both temperature and pressure are kept constant during the cycle.

PRODUCTION OF GLYCEROL

As mentioned in Chapter 4, glycerol is a by-product in the manufacture of soap, and in the splitting of fatty acids.

As produced in the soap plant, glycerol is a component of the spent lye which has been drawn off after the saponification and with the various washes. The glycerol content of the liquor depends on the number of washes used and on the quality of the fats that were saponified. High-grade fats will yield 9–10% of glycerol in the first two lyes drawn off. As each succeeding wash yields less and less glycerol, however, the combined waters that reach the evaporator, may be as low as 3–4% in content of glycerol. When the market is favorable, as much glycerol as possible is extracted from the soap, but when the price is low, part of the glycerol is allowed to remain in the soap and the manufacturer employs only two washes.

In addition to glycerol, spent lye contains soluble soaps, salt, traces of caustic soda, sodium carbonate, sodium silicate (from reboiled scrap soap), organic matter, such as albumin, dirt, coloring matter, and water. All of these extraneous substances must be eliminated before the glycerol can be obtained in its pure form.

If the lye contains much alkali, it may be neutralized by agitating the hot liquor with the requisite amount of fatty acids or house grease, and in this way the alkali can be utilized economically. If

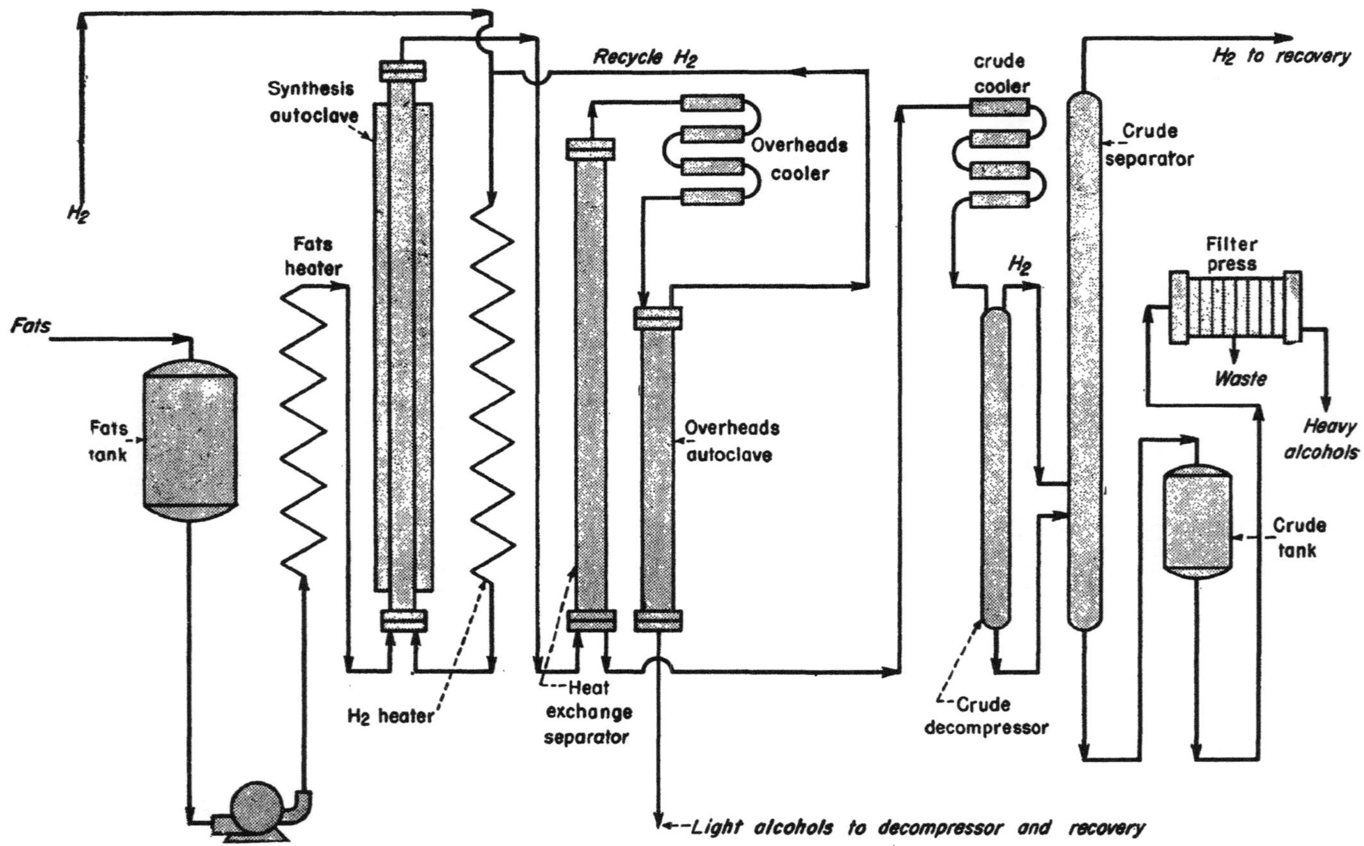

FIG. 6:16. DENORA PROCESS FOR HYDROGENATION OF FATS (Reprinted with permission of *Chemical Engineering*)

little alkali is present, sulfuric or hydrochloric acid may be used as the neutralizer. The spent lye should be left alkaline. Further purification is accomplished by the addition of one-half pound of alum to 1000 pounds of lye; or 2.5 pounds of ferric chloride may be added. The solution is agitated with open steam or compressed air. By this means, soaps and albuminous matter are coagulated and precipitated. In an alternate method, the lye is treated with a strong solution of calcium chloride. This also precipitates impurities and leaves the lye slightly alkaline.

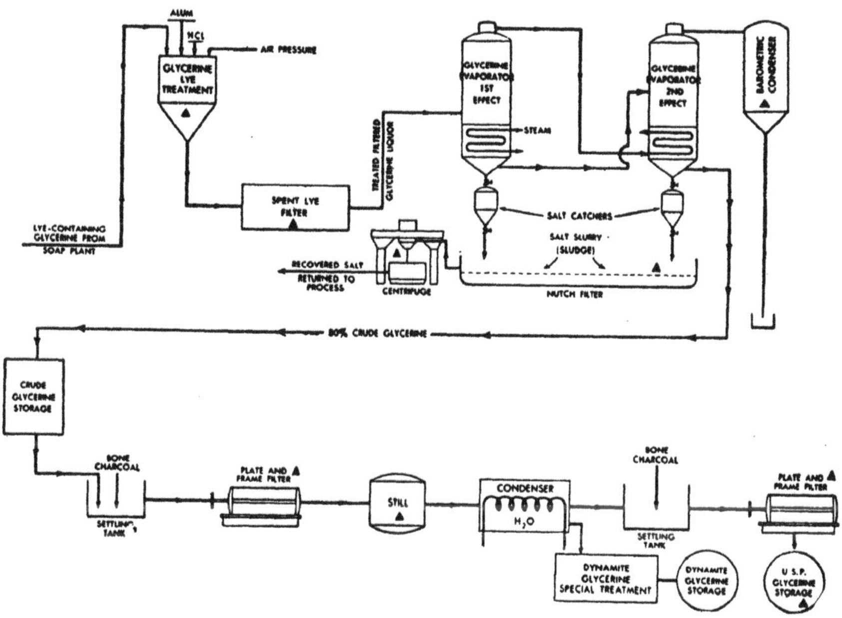

FIG. 6:17. PLAN OF PROCESS FOR RECOVERY OF GLYCEROL
(*International Nickel Co., Inc., New York*)

Where acid or acidic salts are used in the preliminary treatment, the liquor is made slightly alkaline and is filtered. If a test shows that there are still some organic impurities, a second treatment, as before, is given and the liquor is again filtered.

It should be noted that although some manufacturers use sulfuric acid to neutralize the liquor, this procedure is really not a good one. As the result of the reaction between the alkalies and the sulfuric acid, Glauber salt is introduced and this tends to crystallize on the tubes of the evaporator. Figure 6:17 shows a layout for a process for recovery of glycerol. Glauber salt seems to adhere to the tubes to a greater extent than does sodium chloride, resulting in impeded heat transfer. Then too, the fact that the salt recovered from the evaporator is used over in the soap kettles makes the use of sulfuric acid objectionable.

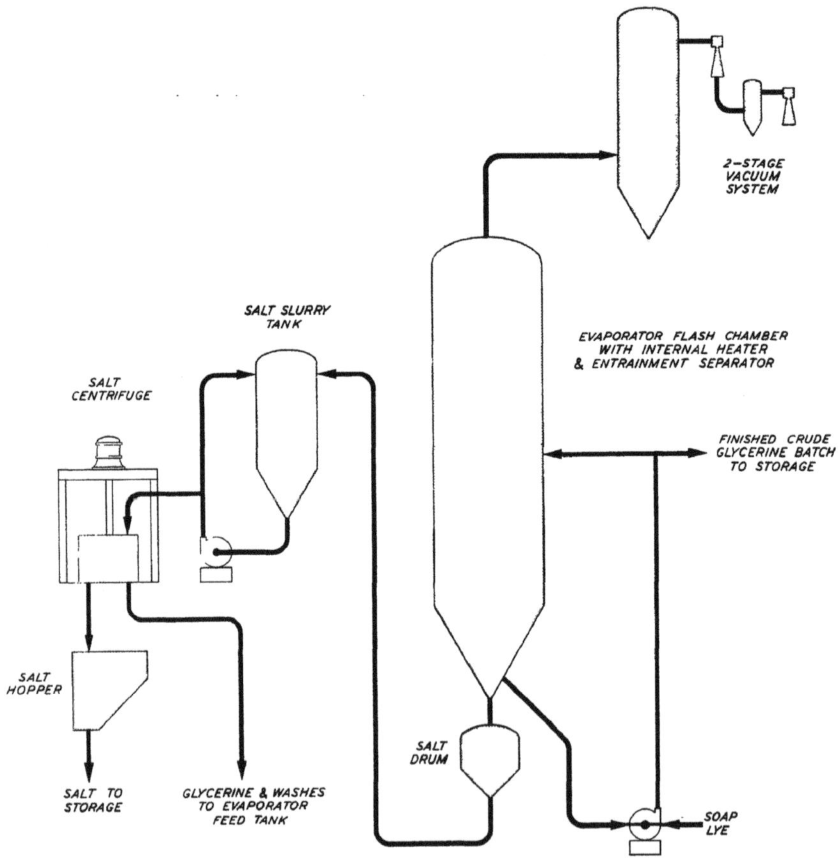

FIG. 6:18. FLOW DIAGRAM FOR EVAPORATION OF GLYCEROL
(*Wurster & Sanger, Inc., Chicago, Ill.*)

The purified liquor is now transferred to a vacuum evaporator (Fig. 6:18). In large-scale production a double-effect evaporator is used, which affords considerable economy in the consumption of steam and water.

As the liquor becomes concentrated, the salt precipitates out. When a concentration of 40% glycerol is reached, the charge known to the trade as "half-crude," is dropped into the salt box, which is situated under the evaporator as part of the vacuum installation. The liquid is drawn off by vacuum through the perforated and canvas-lined bottom of the salt box. The salt is washed with cold water to remove occluded glycerol and the liquid half-crude is further evaporated until it attains a density of 33.5°Bé, which indicates about 80% glycerol. The concentrate is dropped into the salt box and the glycerol is filtered off. The salt is then washed with cold water as nearly free of glycerol as possible, steamed to dry it somewhat, and returned to the kettle room to be used again.

The crude is now ready to be distilled for *dynamite grade*, which is yellowish and has a specific gravity of 1.262, or it may be bleached with activated carbon and reduced with water, to chemically pure (C.P.) grade which has a specific gravity of 1.249 at 25°C and contains about 96% of glycerol.

Crude glycerol boils at 160°F under 27 inches of vacuum, in spite of its 80% glycerol content and the salt that is dissolved in it. Under atmospheric pressure, C.P. glycerol boils at 554°F with some decomposition. From this it is obvious that vacuum distillation is more economical, for it saves fuel and loss of glycerol through carry-over with the vapor.

Glycerol from Alcohols

The process adopted by DuPont, which utilizes methanol or ethanol, results in the formation of the methyl or ethyl esters and glycerol. The reaction is:

$$
\begin{array}{ccccccc}
C_{17}H_{35}COOCH_2 & & CH_3OH & & C_{17}H_{35}COOCH_3 & & CH_2OH \\
| & & & & & & | \\
C_{17}H_{35}COOCH & + & CH_3OH & \rightarrow & C_{17}H_{35}COOCH_3 & + & CH\ OH \\
| & & & & & & | \\
C_{17}H_{35}COOCH_2 & & CH_3OH & & C_{17}H_{35}COOCH_3 & & CH_2OH \\
\text{Glyceryl Stearate} & & \text{Methanol} & & \text{Methyl stearate} & & \text{Glycerol} \\
\text{(Tallow)} & & & & & &
\end{array}
$$

This reaction works best with neutral fats, containing no free fatty acids, and catalyzed with caustic soda. Based on the weight of glyceride to be treated, 0.1–0.5% of caustic soda is dissolved in an amount of anhydrous methanol—not over 1.75 equivalents of the glyceride.[26]

For example, 1000 parts of refined coconut oil together with a solution of 5 parts of NaOH dissolved in 240 parts of anhydrous methanol are heated to 80°C and stirred for 15 minutes. The batch is allowed to stand. About 90% of the entire glycerol content of the oil separates out at the bottom of the mixer, and may be drawn off, as 60–70% crude glycerol, for distillation of pure glycerol. The esters may be washed once or twice with water, to extract the remaining glycerol. The esters may then be combined with the proper amount of caustic soda to form soap and alcohol, which is regenerated, and may be distilled off leaving practically anhydrous soap.

This is a very easy method for producing crude glycerol, because the process does not require special equipment, except for flashing off the alcohol.

Saponification Crude

Crude glycerol produced by the fat-splitting process is purer than that obtained from soap lye. Most of the impurities remain with the fatty acids and, therefore, the "sweet-waters" resulting from the hydrolysis contains sulfuric acid as the principal impurity and water. Its treatment consists of neutralizing with lime. After filtering, the liquor is evaporated to a concentration of 88% glycerol, and is known in commerce as *88% saponification crude*,* which requires only distillation.

Where the production of crude is too small for continuous operation of a glycerol distillery, the crude may be sold to larger producers. Where a plant is located in close proximity to a larger one, arrangements may be made to sell spent lye or sweet-waters. In this way, the cost of installing a glycerol plant may be avoided entirely.

Synthetic Glycerol

There are at present at least five companies manufacturing synthetic glycerol in the United States. Their capacity per annum in million pounds is:

Shell Chemical Corp. (Houston Tex.)	110
Shell Chemical Corp., Norco, La.	35
Dow Chemical Co., Freeport, Tex.	70
Olin-Mathieson Chem. Corp., Brandenburg, Ky.	40
Atlas Chemical Ind., Atlas Pt., Del.	25
Total	280

The Shell process consists of six steps:

* The term *saponification*, as applied to crude glycerol obtained from the fat-splitting process, is derived from the old nomenclature, in which hydrolysis was called *acid saponification* of fats.

1. Chlorination of propylene at 500°C to produce allyl chloride. (Propylene is one of the gaseous products of the cracking process.)
2. Hydrolysis of allyl chloride with caustic soda.
3. Chlorinhydration of allyl alcohol with a solution of hypochlorous acid, produced by the reaction of chlorine with caustic soda.
4. Treatment of glycerol monochlorohydrin with 10% of caustic soda and 1% of sodium carbonate, thus converting it to glycerol.
5. Extraction of the glycerol with xylene.
6. Distillation *in vacuo*

The synthetic glycerol produced has been tested and found to be equal to natural grade C.P.

In a process for production of glycerol, used by E. I. du Pont de Nemours & Co., carbon monoxide is condensed with formaldehyde to yield first glycolic acid, and after further treatment, glycerol.[27]

The Office of Technical Services, U. S. Department of Commerce released a report dated Aug. 13, 1946, describing a process by which the I. G. Farben plant at Hoechst, Germany, was able to produce 180 tons a month of a synthetic glycerol composition. In this process, sugar was first treated in an aqueous solution with 0.1% oxalic acid at 70 to 80°C. The solution was then neutralized with calcium carbonate. Reduction was carried out continuously in a vertical, heated return-pipe assembly, using nickel deposited on pumice as a catalyst. The time of contact with the catalyst was about 8 hours and the reaction was carried out at 200°C and 300 atmospheres pressure. After the reaction was complete, the solids were allowed to settle. The solution was then treated with activated charcoal and distilled. The resulting product contained 40% glycerol, 40% propylene glycol and 20% alcohols containing six carbon atoms. The yield was about 95%. The purified glycerol was reported to be as good as that obtained from fats.

The Atlas process is also based on sugar.

MISCELLANEOUS

Oxidation

One of the problems in industry is the control of the deleterious effects of oxidation. Everyone is aware of the spoilage of foods such as oils, fats, meats, potato chips, peanuts, candy, and coffee. Organic coloring matter, petroleum oils, and rubber are other materials that deteriorate through oxidation.

Heat, air, light, and moisture favor oxidation, which is sometimes accompanied by rancidity. Enzymes from seed pulp and fatty tissue, bacteria, and various microorganisms are also responsible for rancidity. This phenomenon is also hastened by the presence of metallic catalysts, such as copper and iron.

It is well recognized that unsaturated compounds yield more readily to oxidation than do the saturated. For example tung and linseed oils, both of which contain highly unsaturated glycerides, are the best oxidizing oils, and for this reason they are used in paint and varnish where oxidation and drying are necessary. On the other hand, coconut oil and tallow do not readily oxidize. These two fats form the basis of most toilet soaps, which remain white for a long period.

Not only do oils and fats oxidize and become rancid and discolored, but these effects continue even after saponification, so that oxidative influences must be prevented even in soap.

Although, in most cases of rancidity, the content of free fatty acids parallels the degree of rancidity, the fact that an oil is rancid does not always mean that it is high in free fatty acids. The rancidity of fat results from changes that give rise to malodorous substances. The products formed in rancid fat are mixtures of aldehydes, ketones, lactones, oxy- and hydroxy-fatty acids, alcohols, carbon dioxide, and moisture. Such products have been isolated as azelaic, pelargonic, and heptylic aldehydes and acetic and nonylic acids and their aldehydes, according to Jamieson.[28] Most of these compounds have characteristic and unpleasant odors.

One of the best known tests for rancidity, is the Kreis test (see Chapter 9). Considering the sensitivity of this test, Jamieson interprets the results by placing oils and fats into four categories:

1. Those giving no reaction; these will withstand severe exposure before turning rancid.
2. Those giving a reaction when diluted, represent products which have not turned rancid as far as odor and taste are concerned, but in which changes are already in progress which will later result in rancidity.
3. Those which give a reaction when diluted 1:10, but none in a dilution of 1:20, represent a late stage of incipient rancidity, generally evident by bad taste and odor.
4. Those which give a reaction in a dilution of 1:20 show definite rancidity. Exception is crude cottonseed oil, which gives a definite Kreis reaction, although free from rancidity.

Antioxidants

Many compounds have been patented for delay of oxidation. Antioxidants or inhibitors used in industrial products are of great variety. Those used for foods must be odorless, tasteless, and non-toxic and must not produce undesirable color. Furthermore they must be approved by the Food and Drug Administration.

For oleomargarine, mayonnaise, salad dressings, confections, and baked goods, the antioxidant properties of lecithin are helpful.

Mannitol triacetate is recommended as a potential antioxidant and color stabilizer for vegetable oils.

According to British patent 492 714, vegetable and essential oils, fats, and coloring matter are preserved against oxidation by incorporating moderately roasted cocoa beans or roasted cocoa-bean shells; or by treating the oils with an aqueous extract so as to absorb the antioxidants.

Gum guaiac, added to lard (0.05%) preserves it and even extends its protection to the fat after use in baking.

Nordihydroguaiaretic acid (NDGA), in concentrations of 0.002–0.005% with suitable synergists, has been recommended by A. Foyno. According to Golumbic, α-tocopherol and 1,4-naphthol hydroquinone have antioxidant properties. A Canadian patent covers the use of 0.001–3% of a partial carboxylic acid ester of gum guaiac.

Propyl gallate ** is employed as a stabilizer for compounds subject to autoxidation. Its producers say that it effectively inhibits formation of peroxides in fats, fatty oils, fatty acids, and other substances. Golumbic and Mattill found that ascorbic acid enhances the antioxidant properties of tocopherols,[29] hydroxychromans, hydroquinone, and related compounds.

Butylated hydroxyanisol (BHA)*** is now used in petroleum products, butylated hydroxytoluene (BHT) is widely used in animal feeds, BHT and BHA and propyl gallate, but not BHT and propyl gallate are often synergistic (that is, they reinforce one another in their antioxidant properties). BHA and BHT, alone or mixed, are excellent for cosmetics.

An order on food additives was issued by the Food and Drug Administration, clearing the use of THBP (Eastman Kodak's 2,4,5-trihydroxybutyrophenone) as an antioxidant in food, alone or in combination with other permitted antioxidants, up to 0.02% of the

---

Suppliers: * 158; ** 68; *** 49.

oil or fat content of the food. This approval was granted under the *Food Additive Regulation* 21, CFR 121,1116 [30] as a food additive at the 0.02% level. This is the same type of approval granted for BHA and BHT.

Hydrogen Ion Concentration

The control of processes by *hydrogen ion concentration*, or *p*H, is well established in the chemical industry. As *p*H is frequently mentioned throughout the book, this brief discussion should be sufficient to inform those readers who have as yet not been initiated to the term.

TABLE 6:4

*p*H OF 1/10 SOLUTIONS OF FAMILIAR ACIDS

Acid	*pH*
Hydrochloric	1.0
Sulfuric	1.2
Phosphoric	1.5
Acetic	2.9
Boric	5.2

Sorensen, the great Danish chemist, proposed the method for expressing hydrogen ion concentration as it is now in general use. For a neutral solution, the concentration of hydrogen, is 1 part in 10 million. As this is somewhat awkward to apply, he proposed that the logarithm of its reciprocal be used, i.e., 7. Distilled water has a *p*H of 7 at 22°C. Acid solutions vary in strength increasingly from *p*H 7 to 0. Alkaline solutions increase in strength from *p*H 7 to 14. It

TABLE 6:5

*p*H OF 1% SOLUTIONS OF FAMILIAR CHEMICALS

Caustic Soda	13.3
Sodium Metasilicate	12.4
Trisodium Phosphate 12 H_2O	12.0
Tetrasodium pyrophosphate	10.2
Tetrapotassium pyrophosphate	10.2
Sodium Tripolyphosphate	9.7
Borax	9.0
Disodium Phosphate	8.8
Sodium Tetraphosphate	8.5
Sodium Hexametaphosphate	6.8
Sodium Fluoride	6.5
Monosodium Phosphate	4.8
Sodium Acid Pyrophosphate	4.2
Sodium Silicofluoride (Saturated solution)	3.5

should be noted also, that the *p*H of solutions increases slightly as the temperature is elevated.

Tables 6:4 and 6:5 give the *p*H of several familiar chemicals. Dilution of these chemicals does not materially alter the *p*H.

From Table 6:5 it is obvious that some salts give acidic reactions. Besides those shown, potassium alum, stannous chloride, and others might be cited. By mixing an alkali (or an acid) of definite *p*H with an acid or acid salt, it is possible to obtain a solution of any *p*H required. Certain salts may be added to solutions to stabilize the *p*H. These are called *buffers*. Phosphates, borates, phthalates, and carbonates can act as buffers.

Uses of *p*H Determination

Determination of *p*H is very useful in the preparation of emulsions because there is an optimum *p*H at which every emulsion is stable. For example, the chief ingredients of mayonnaise are oil, egg yolk, and water. It has been found that the addition of vinegar (*p*H 2.9) stabilizes the emulsion. Other emulsions require alkaline reagents to stabilize them. For example, soap (*p*H 9–10) is used in the preparation of kerosine emulsions for horticultural sprays. Likewise, soap stabilizes cutting oils.

Whereas the so-called soda and potash soaps have a *p*H of 9–10, the soaps produced with triethanolamine and other amines have a *p*H slightly over 8. The *p*H of the emulsifier is of great importance in such products as cosmetics and medicinal preparations, where the emulsion is to be applied to tender skin.

The *p*H is also of prime importance in the case of sulfonated oils. If the *p*H of the oil is under 7, it is of moderate fluidity and good clarity, whereas an oil that is made too alkaline is thick and cloudy.

*p*H control is useful in the manufacture of water-emulsion waxes in which combinations of various vegetable and mineral waxes are used in conjunction with resins to form the emulsions.

Measurement of *p*H

The most accurate means of determining the *p*H of a solution is the potentiometer. This instrument is made in many forms depending on its purpose. For work requiring great precision there is a type that is accurate to 0.001 *p*H. The *p*H recorder is an instrument that furnishes a continuous record of acidity or alkalinity of a solution. Such a record may be desirable in a process such as wool scouring, where in a continuous process, starting with an alkaline soapy solution, there is a gradual diminution in the alkaline strength of the bath. The continuous record of the *p*H will indicate when further additions of soap and alkali are necessary in order to maintain proper scouring.

To simplify this determination for persons without technical training, there are several devices on the market. They depend on color reagents which indicate a certain range of *p*H (see Table 6:6).

TABLE 6:6

*p*H RANGE AND COLOR CHANGE OF INDICATORS
(Prepared by W. A. Taylor Co.)

Indicator	*pH Range*	*Color Change*
Acid Cresol Red	0.2–1.8	Red–Yellow
Thymol Blue	1.2–2.8	Red–Yellow
n-Dinitrophenol	2.8–4.4	Colorless–Deep Yellow
Bromcresol Green	3.8–5.4	Yellow–Blue
Methyl Red	4.4–6.0	Red–Yellow
Bromthymol Blue	6.0–7.6	Yellow–Blue
Cresol Red	7.2–8.8	Yellow–Red
Thymol Blue	8.0–9.6	Yellow–Blue
Phthalein Blue	8.8–10.4	Yellow–Blue
Tolyl Red	9.6–11.2	Purple–Red
Parazo Orange	11.0–12.6	Yellow–Orange
Acyl Blue	12.0–13.6	Red–Yellow

Standardized *p*H Sets

In most of these testing instruments, tubes or ampules containing buffered solutions, accurately standardized for a definite range of *p*H, are used in a block, called a *comparator*. To an equal volume of unknown solution, 0.5 ml of the indicator for the expected range is added. It is then compared with the standard that matches it most closely.

a) The *LaMotte* comparator consists of sets of sealed vials or standards, which are placed in a block and used to match an unknown containing the same color reagent as the standard.

b) The *Taylor* comparator is similar to the above, but the standards are set into slides made of Bakelite. Therefore the ampules need not be handled individually. This provides several standards with a greater range, at each trial of a single slide. It also avoids the possibility of breakage.

c) The *Hellige* comparator, in place of vials, employs discs containing small colored glass windows. The disc of the approximate *p*H is set into a block containing graduated test tubes of the solution to be tested. According to the manufacturer of the instrument, readings can be made that are accurate to 0.1 *p*H.

For rapid tests, rolls or strips of *p*H paper may be purchased. These are immersed in the liquid to be tested and they are then compared with colors on the container or roll dispenser.

There is also a universal indicator solution for use in determining approximate *p*H.

"Parstains" are indicators put up in capsules for ready use.

Hardness of Water

Whenever water is discussed, whether it be for chemical manufacturing, or for laundering and textile processing, there is almost invariably an allusion to its hardness.

Water is never found absolutely pure in nature. As we all know, it dissolves almost every variety of mineral and some organic matter too. The solvent action of water on rock is best illustrated by such phenomena as crystal caves which are well distributed over this country and other parts of the world. These caves were formed by erosion and subsequent deposition of limestone which is the material of the stalagmites and stalactites. Natural waters may also carry organic matter in solution from decayed vegetation and seepage from human and animal excreta. Besides being in solution, impurities may exist as suspended matter, such as silt which is carried down a river.

The purest form of water to be found in nature is rain water. Even as such, it contains traces of ammonia, and other gases which it has picked up from the atmosphere. Rivers, springs, lakes, all contain matter which the water has dissolved during its course on the surface or through the ground. Therefore we can expect to find dissolved in the water, nitrates, and nitrites from decomposed nitrogenous matter, carbon dioxide, oxygen, sodium chloride and sodium sulfate, bicarbonate and carbonates of calcium (limestone), calcium and magnesium sulfates and chlorides, iron and aluminum salts, all depending on the kind of strata the water has traversed. (See Fig. 6:19.)

Sodium salts, in small quantities, are not detrimental in soap solutions such as may be used in laundering and textile processing, or in soapmaking. Other metallic salts, however, have a decidedly deleterious effect because of their reaction with soap to form insoluble compounds. It is the proportion of these metallic salts contained in water that determines its hardness, the chief offenders being the ever-present salts of calcium and magnesium.

According to American engineering practice, one degree of hardness means that in one U.S. gallon of water, weighing 58,381 grains, the amount of hardness-forming or soap-destroying chemicals is equivalent to one grain of calcium carbonate, i.e., 1 part in 58,381 parts.

Temporary hardness is caused by calcium bicarbonate and may be overcome by boiling, the reaction being:

$$Ca(HCO_3)_2 + \text{heat} = CaCO_3 \downarrow + CO_2 + H_2O$$

This reaction is not complete, for there is some solubility of the calcium carbonate. The CO_2 passes off with the steam. The residual or permanent hardness is caused by the sulfates of both calcium and magnesium. Magnesium bicarbonate, even though it is converted by heat to magnesium carbonate, is soluble in hot water, and in a steam boiler it may become concentrated and cemented with sulfate and other particles to form a hard, resistant scale.

Water containing 0 to 3 grains per gallon is called *soft water*; from 4 to 7 grains, *moderately hard*; from 7 to 12 grains, *hard*; beyond this, the water is considered very hard.

One grain of hardness, in terms of the U.S. gallon, is equivalent to 1.20 grain in the Imperial gallon (British). Figured as parts per million, they are thus, respectively, 17.1 and 14.3.

Water that contains over 3 grains of hardness, should be softened for use in the manufacture of liquid soap, in laundering, textile processing, and for making steam.

What happens when soap is dissolved in hard water? Immediately upon dissolution, a turbidity or curdling effect is noticeable. This is due to the insoluble lime and magnesia soaps formed. These metallic soaps are sticky and become affixed to the fabrics in washing. Metallic soaps are manufactured on a large scale for water-proofing compounds and for lubricants. It is therefore quite evident why such material affixed to the clothes, impede the washing process.

The National Laundryowners' Association reports that water of 3.7 grains hardness, cold, without the addition of alkali, decomposed 1.89 pound of 88% soap per 1000 gallons. Pure distilled water consumed 4 ounces per 1000 gallons. When 0.11 grain of soda ash per gallon was added to the same hard, hot water, only 12 ounces of soap was consumed. It is obvious therefore, that much soap may be saved by preliminary softening of hard water. Treatment of hard water results in economy and also in improved washing, by avoiding deposition of insoluble soaps on the fabric, or other surfaces.

Water-softening equipment utilizing natural or synthetic zeolites is used to a great extent in industry. When the zeolites have reached the limits of their capacity to soften the water they are regenerated with salt. Another means of softening water and even further purifying it, so that it is on par with distilled water is by means of ionization equipment.

Sequestering (Chelating) Agents

The presence of calcium, magnesium, and iron compounds in most water supplies, causes the precipitation and clouding of soaps. The precipitated metallic soaps are especially offensive in laundering because they result in stains when the fabric is ironed. This precipitation and consequent turbidity is especially evident in liquid soaps.

In the absence of water that has been softened by means of zeolite, deionization, or distillation, organic *sequestering* or *chelating* agents are used, wherever contamination from metallic ions has a detrimental effect on manufacturing processes or formulated products.

According to John J. Singer of the Hampshire Chemical Corp., Nashua, N.H., the role of amino acid chelates is differentiated from that of the polymerized phosphates, which are really inorganic chelating agents. In practical operation, the differences are few, but they are similar in many respects.[31]

1. Amino acid chelates combine with most every polyvalent metal ion, whereas the polyphosphates effectively complex only calcium and magnesium.
2. The polyphosphates under certain conditions revert to orthophosphates at various rates depending on *p*H and temperature (mostly hot), with resultant loss of sequestering power.
3. The polyphosphates have the very valuable *threshold effect*, while the amino acids and their salts do not.*
4. In the past, amino acid chelating agents have been more expensive to use than the polyphosphates, for water softening purposes. There are now amino acid chelating agents which compete fairly with the polyphosphates. These modify crystal growth of precipitates, not as polyphosphates do, but in other ways, which often supplement the threshold effect.

The two types of chelating agents differ as follows:

---

* The *threshold effect* is the ability of polymerized phosphates to temporarily inhibit the precipitation of insoluble materials, even though only a few ppm of the polyphosphate are present. The effect may last for from only a few minutes to many hours, depending on the system.

FUNCTION	PERFORMANCE
Solubilize and disperse proteins	Amino acid type solubilizes; polyphosphates disperse
Prevent rancidity by inactivating pro-oxidant metals	Only the amino acids types perform this way
Protect coloring or perfuming agents	Amino acids type chelate
Stabilize bleach components	,, ,, ,, ,,
Dissolve scale or other deposits not ordinarily susceptible to detergent ingredients alone	,, ,, ,, ,,
Increase shelf life of liquid formulations	,, ,, ,, ,,

Both types of chelating agents perform similar functions as follows:

1. Solubilize or prevent the precipitation of heavy metal soaps
2. Build or boost cleaning action of detergents
3. Stabilize or increase foam.
4. Combine with calcium and magnesium to soften water
5. Reduce adsorption of detergent ingredients on surface cleaned
6. Lower ash content of textile fibers
7. Promote free rinsing
8. Prevent or minimize calcium and magnesium salts of some synthetic detergents
9. Protect and make quaternary phenolic and other bactericides more effective
10. Prevent build-up of scale in industrial cleaning equipment
11. Increase efficiency of optical bleaches.

The polyphosphates alone delay the formation of precipitates at low concentrations ("threshold effect").

Nitrilotriacetic acid (NTA) is so similar to EDTA that the compounds are almost chemical twins. The overwhelming advantage of NTA is its low molecular weight. Only 2.75 parts of commercially available NTA-Na_3 are required to complex 1 part of $CaCO_3$, whereas 4.16 parts of EDTA dihydrate are required for 1 part of $CaCO_3$. This makes NTA at least 1.5 times as effective, pound for pound, as EDTA, at its theoretical maximum. NTA does not cost more than EDTA. One pound of NTA is capable of softening the same amount of water as can be softened by 5–6 pounds of tripolyphosphate. In soap, however, NTA is not more effective than EDTA because of the great precipitating power of the fatty acids for calcium and magnesium.

The organic chelating agents are sold on a percent-active basis. Some users find the solutions easier to handle; others prefer the dry powdered concentrated forms, which contain up to 100% of active ingredient.

TABLE 6:7

ORGANIC CHELATING AGENTS

Product	*Composition*	*Source*
Chelon		Cowles
Cheelox	Aminocarboxylic acid salts	Antara
Chell 85	Sodium dihydroxy ethylglycine	Geigy
Cyquest 40	40% Sodium salt of EDTA	Cyanamid
NTA	Nitrilotriacetic acid salts	Hampshire
Nullapon	Same as Cheelox	Antara
Perma-Kleer	Sodium salt of polyaminocarboxylic acid	Refined Products
—	Sodium Gluconate, gluconic acid	Pfizer, Cowles, Industrial Chemicals
Sealene Fe 900,	35%, 50% Solid content	Staley
Sequestrene AA	100% powdered EDTA	Geigy
Tetrene	EDTA and salts	Glyco
Versene, Versenol	EDTA and salts	Dow

It has been our practice to use about 1/10% of the concentrated powder, based on the weight of material to be chelated. Thus, 100 gallons of liquid soap would require 1 pound of reagent. If liquid reagents are used, the proportion to use will be in accordance with the percent concentration of active chelate.

Table 6:7 lists some of the prominent organic chelating agents.

REFERENCES

1. USP: 2 726 253 (1955); 2 727 814 (1955); 2 729 661 (1956); 2 734 635 (1956); 2 759 956 (1956).
2. E. L. D'Aquin *et al.*, USP 2 857 411 (1958).
3. *Ind. Eng. Chem.*, 49 (6):922 (1957).
4. *Chemical Week*, 11/9/1963.
5. S. W. Gloyer, *Ind. Eng. Chem.*, 40:228 (1948).
6. *Oleagineux*, 3:532 (1948).
7. T. P. Hilditch, Distillation, in *Industrial Chemistry of Fats and Waxes*; D. Van Nostrand Co., Inc., New York, 1941.
8. J. Mattiello, *Protective and Decorative Coatings*; Government Printing Office, Washington, D.C., 1945.
9. *Ind. Eng. Chem.*, 32:10 (1940).
10. A. Schwarcman, USP 2 140 271.
11. J. Scheiber, Ger P 513 309; USP 1 942 778.
12. H. Ufer, USP 1 892 529.
13. N. T. Joyner, *The Plasticizing of Edible Fats*; Votator Div., Chemetron Corp., Louisville, Ky.

14. *J. Am. Oil Chemists Assoc.*, 30(11):526 (1953).
15. Brit P 1515 (1903).
16. Ger P 141 029 (1903).
17. USP: 1 004 035; 1 008 474 (1911).
18. USP: 1 628 190 (1927).
19. USP: 1 915 473.
20. USP: 2 139 602.
21. *Sulfation of Fatty Products:* Bull. 203C; Emery Industries.
22. USP: 2 149 397; 2 156 863; 2 271 619.
23. Wurster and Sanger: Personal communication (1965).
24. USP: 1 622 126; 1 766 863; 1 951 241; Brit P 277 085.
25. USP: 2 224 925; 2 224 984; R. H. Potts and J. E. McKee, assigned to Armour and Co.
26. USP: 2 271 619; Bradshaw and Meuly; assigned to Du Pont.
27. USP: 2 211 626; E. I. du Pont de Nemours.
28. G. J. Jamieson, *Vegetable Fats and Oils*; Reinhold Publishing Corp, New York, 1943.
29. C. Golumbic and H. A. Mattill, *J. Am. Chem. Soc.*, 63:1279 (1941).
30. *Federal Register*, 28:1141 (1963).
31. John J. Singer, Chelating Agents in Detergents, *Soap Chem. Specialties*, October 1961.

chapter 7

SOAP MANUFACTURE

Location of Plant

A soap plant should be located on the outskirts of a town or city to avoid objectionable odors in residential sections. A desirable location is on the railroad to facilitate the transport of raw materials to the plant and finished products out of the factory. In all the large cities in the United States there is door-to-door service of trailer loads and tank trucks for liquids. Proximity to water is also desirable so as to furnish cheap water supply for plant needs and to discharge biodegradable effluents from plant operations. (Some effluents must be treated before discharging them into a river.)

For a small plant, a one-story building with a high ceiling and adequate light will furnish proper housing. Sufficient yard space should be provided for spur track, for storing drums and materials that do not require inside storage, and for parking cars.

Equipment

The best source of heat is the steam boiler. By this means, heat can be supplied anywhere on the premises to kettles, crutchers, tanks, and pumps.

If the plant is small, or if steam is used intermittently, an oil-fired steam boiler operating at 60–75 pounds of boiler pressure is adequate.

Kettles:

The size of kettles should conform to the desired production. If the plant is to produce boiled soaps on a small scale, one or two kettles of 10,000-pound capacity may be installed. The kettles should be sufficiently elevated to permit movement of 1200-pound frames beneath them.

The kettles are equipped with both open (perforated) and closed steam coils. Round kettles with conical bottoms are preferred by most soapmakers. Each kettle should be equipped with a valve at the bottom and a draw-off pipe which is jointed, so that it can be drawn up and down (Fig. 7:1).

The pumps for pumping the hot soap may be rotary electric pumps or piston-type steam pumps, connected to the draw-pipe from the

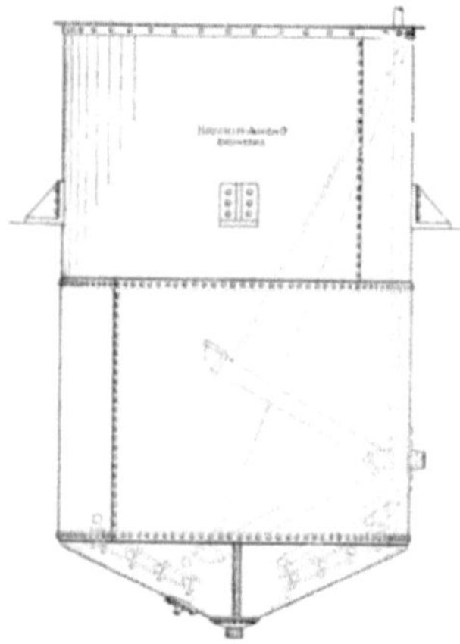

FIG. 7:1. KETTLE FOR BOILING SOAP (Houchin)
(*Benjamin Eastwood Co., Paterson, N.J.*)

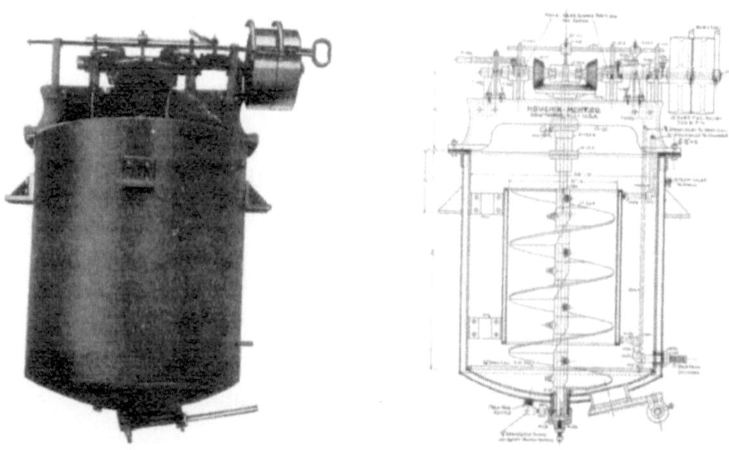

a *b*

FIG. 7:2. PERFECTION JACKETED CRUTCHER (Houchin)
(*Benjamin Eastwood Co., Paterson, N.J.*)

FIG. 7:3. STANDARD SOAP FRAME (Houchin)
(*Benjamin Eastwood Co., Paterson, N.J.*)

kettle, which should be about 2.5 inches inside diameter. Valves and pumps should be so installed as to allow steaming out of all pipe lines before and after use. Hot soap, remaining in the pipe line will congeal and clog.

Crutcher:

The crutcher is another piece of standard equipment in the soap-plant. It is a steam-jacketed tank with an agitator (Fig. 7:2*a*). Its

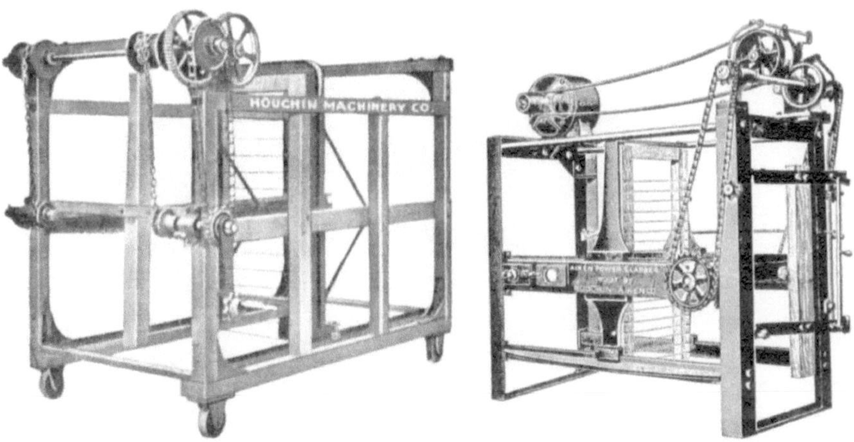

FIG. 7:4. SLABBERS (Houchin)
(*Benjamin Eastwood Co., Paterson, N.J.*)

function will be revealed in the discussion of the manufacturing process. A preferred type of crutcher is one with a worm-type agitator, surrounded by a steam-jacketed cylinder (Fig. 7:2*b*). The drive is geared so as to provide clockwise and counter-clock wise motion. A 1200-pound crutcher may be run by a 5-hp motor.

Frame:

For the manufacture of bar soap, about twenty soap frames of 1200-lb capacity are necessary. The frame in Figure 7:3 is a truck with collapsible steel sides. The hot soap from the crutcher is run into the frames where it is allowed to cool and congeal.

Slabber:

A *slabber* (Fig. 7:4) is a machine with wires strung horizontally.

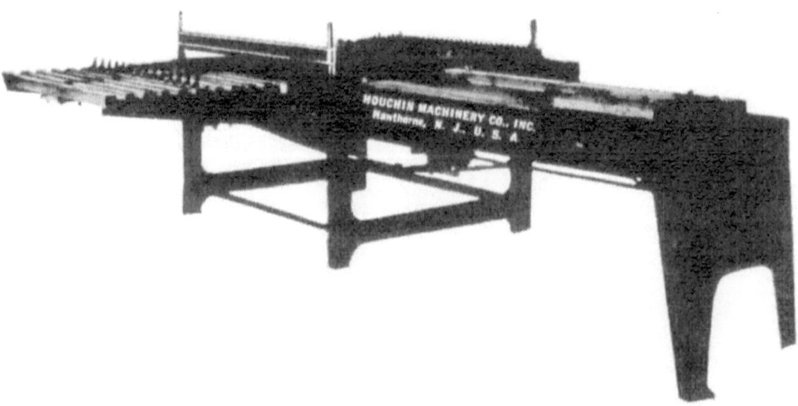

FIG. 7:5. TYPES OF CUTTING TABLES (Houchin)
Upper: Manually operated; lower: power-driven
(*Benjamin Eastwood Co., Paterson, N.J.*)

This is used to cut the 1200-lb block of soap into slabs as thick as the individual bar of finished soap.

Cutting Table:

The *cutting table* (Fig. 7:5) is a machine that cuts the slab into bars. It also places the bars on the racks for drying before pressing. It may be either hand- or power-driven.

FIG. 7:6. STANDARD SOAP RACK AND TRUCK (Houchin)

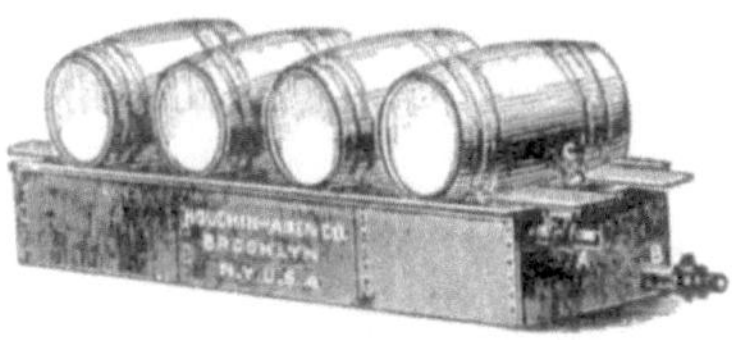

FIG. 7:7. STEAMING-OUT TANK (Houchin)

FIG. 7:8. PRESS FOR CAKES OF SOAP

(All from Benjamin Eastwood Co., Paterson, N.J.)

Miscellaneous:

About 100 *racks* (Fig. 7:6) should be provided.

For manufacture on a small scale, drums of tallow and oils may be placed on rails welded to the top of the kettle. Steam is run into the drums by means of a flexible metal hose. Thus, the fat is steamed out directly into the kettle and the condensed water is used in the saponification. An alternate method is to employ the standard *steaming-out tank* (Fig. 7:7) set directly over a storage tank with a capacity for the entire charge of fat, with condensed water. At the end of the steaming-out operation the water is drawn off and the clean oil may be pumped wherever necessary.

Miscellaneous tanks for storage of oils, alkalies, etc., will be required. Alkali-storage tanks should be set above the kettles, so that the alkali may be fed by gravity. Platform scales of various sizes will be needed at certain points in the plant. A most desirable position for a scale is near the receiving door, so that all materials entering the plant may be weighed and immediately compared with invoices.

Although bar soaps will be sufficiently dry overnight for pressing (Fig. 7:8) and wrapping, in order to facilitate drying, a *tunnel dryer* is used. This is a steam-heated drying closet with fans to circulate the air. Such a dryer will accommodate several truckloads of racks.

Using both kettles and auxiliary equipment, the plant should have a weekly capacity of 15,000 pounds of soap.

PROCESS OF SAPONIFICATION

Saponification—literally, *soap-making*—is the reaction between a fat and an alkali. Whenever any saponifiable oil or fat is treated with a solution of caustic soda or potash, a soap is formed. Soap really consists of the sodium or potassium salts of the fatty acids contained in the oil or fat used. It may be considered chemically as a mixture of sodium or potassium stearate, oleate, laurate, and palmitate; or of salts of other fatty acids, depending on the type of fat used.

The saponification reaction between glyceryl stearate and caustic soda may be represented as follows:

$$
\begin{array}{lclclcl}
C_{17}H_{35}COO\,CH_2 & + & NaOH & \rightarrow & C_{17}H_{35}COONa & + & CH_2OH \\
\qquad\qquad\quad | & & & & & & | \\
C_{17}H_{35}COO\,CH & + & NaOH & \rightarrow & C_{17}H_{35}COONa & + & CHOH \\
\qquad\qquad\quad | & & & & & & | \\
C_{17}H_{35}COO\,CH_2 & + & NaOH & \rightarrow & C_{17}H_{35}COONa & + & CH_2OH \\
\text{Glyceryl Tristearate} & & \text{Caustic Soda} & & \text{Sodium Stearate} & & \text{Glycerol}
\end{array}
$$

This means that one molecule of tallow assuming it to be glyceryl tristearate (which is only one of its components), having a molecular weight of 890, combined with three molecules of caustic soda, with a total molecular weight of 120. In other words, they combine in the approximate ratio of 7 to 1. Therefore, to saponify 100 pounds of tallow, 14 pounds of 100% caustic soda will be required. For potash soap, the ratio is 890:168; that is 17 pounds of 100% caustic potash will be needed for 100 pounds of tallow.

Saponification Number

Table 3:4 in Chapter 3 shows the saponification numbers of the various oils and fats. These numbers represent the number of milligrams of 100% caustic potash required to saponify 1 gram of oil. In order to calculate the amount of caustic soda necessary, we must compare the relative molecular weights of caustic potash (56) and caustic soda (40). The saponification number multiplied by 40/56 will give the required amount of caustic soda. In other words, if the saponification number of tallow is 196, then

$$196 \times 40/56 = 140 \text{ mg caustic soda}$$

will be necessary to saponify 1 gram of tallow; or 14 pounds of caustic soda will be required for 100 pounds of tallow. This is in agreement with the previous statement.

In practice, 100% alkali is not used nor is commercial alkali actually 100% pure. The soapmaker always prepares an aqueous solution of the alkali, which he allows to cool before using. He uses the Baumé hydrometer to determine the strength of alkaline solutions. (See Baumé Table in the Appendix.)

The saponification reaction always yields glycerol as a by-product. Whenever a neutral oil (glyceride) is saponified by methods known as *boiled*, *half-boiled*, or *cold* processes, glycerol is formed. When soap is made by the cold and half-boiled methods, the glycerol remains in the soap and is not allowed to settle out or be recovered.

BOILED SOAPS

In this process, 3000 pounds of refined tallow and 1000 pounds of Cochin coconut oil are charged into the kettle of 10,000-lb capacity. The stock should be neutral, that is, must not contain free fatty acids. It is good practice to refine both the tallow and the coconut oil before putting it into the kettle. For a very good grade of white soap, the stock should even be bleached, but, if good stock is used, the refining and bleaching are not necessary.

To the fats, 500 pounds of water is added and the boiling is started with both open and closed steam coils. About 50 pounds of 37-Bé caustic soda is added slowly. The strength of the alkali may vary, but for economy in tank space most soapmakers use a fairly strong lye. If 50-Bé caustic soda is being purchased, it may be used as is, adding about 200 more pounds of water to the kettle at the initial stage of boiling.

After boiling for about half an hour, a sample is taken on a trowel and a drop of 1% alcoholic phenolphthalein is added. If there is an intense red coloration immediately, there is still sufficient alkali present and the boiling should be continued until the red coloration appears only after a short interval. At this point it may be safe to place a speck of soap on the tip of the tongue. There should not be an immediate bite, showing that the fat has consumed nearly all the lye added. About 25 lb more of lye may be added and the boiling continued. As more soap is formed in the kettle, the saponification is accelerated. Larger quantities of lye may be added at a time, but each batch of lye added must be absorbed before more is added.

It is good practice to add lye slowly until one becomes thoroughly familiar with the process of boiling soap. If too great an excess of lye is added during the initial stages of boiling, the stock will be "poisoned." This means that it will not saponify or take up the caustic for a long time. The neutral stock absorbs lye very slowly, although, when coconut oil is present, the saponification is somewhat accelerated. The first batch started with neutral fats, without any soap being present, will be the most difficult and, therefore, all of the precautionary measures mentioned should be strictly observed. The saponification of succeeding batches, however, will be facilitated by the presence of *nigre* which is the residue from the previous batch. Some soapmakers add about 20 pounds of salt with the water at the start. This probably keeps the newly formed soap from adhering to the coils and prevents "bunching."

The contents of the kettle rise considerably during the "killing change," as this preliminary saponification is known; therefore the boiling should be carefully controlled so as to avoid overflow. Another sign that must be carefully watched is that of "bunching" wherein, as the saponification nears completion, the whole kettle seems to go into a semi-solid mass, which arrests the smooth rolling of the contents. Salt must be added to thin out the soap, so that it once more becomes more fluid. Water may also be necessary. However, unless the fat has just about taken up all of the calculated amount of lye, salt should be added cautiously, so as not to open up the soap too soon.

The required amount of caustic soda can be calculated as follows:

By reference to Table 3:4, we find that the average saponification number of coconut oil is 260; therefore

$$260 \times 40/56 = 186 \text{ lb of } 100\% \text{ caustic soda}$$

which are necessary to saponify 1000 pounds of coconut oil.

In the same table, the saponification number of tallow is 196; therefore,

$$196 \times 40/56 = 140 \text{ lb of } 100\% \text{ caustic soda}$$

which are necessary to saponify 1000 pounds of tallow.

Furthermore, by reference to the Baumé table, we find that a 37°Bé caustic soda contains 31% of NaOH. The following calculations will show exactly how much lye will be required for the saponification:

$$1000 \text{ lb coconut oil at } \frac{186}{0.31} /1000 \text{ lb} = 600 \text{ lb of } 37°\text{Bé lye}$$

$$3000 \text{ lb tallow at } \frac{140}{0.31} /1000 \text{ lb} = 1355 \text{ lb of } 37°\text{Bé lye}$$

Total 37°-Bé lye required is 1955 pounds.

Consequently, after boiling 2000 pounds of lye with the stock, the saponification should be completed, and the soap should contain a slight excess of alkali.

The boiling is continued for half a day and then about 2% of salt (sodium chloride), based on the contents of the kettle, is added slowly, spreading it over the surface of the soap, with a shovel, while the boiling is continued. The soap, which has been a smooth rolling liquid heretofore, takes on a rough appearance and begins to grain out. This indicates that sufficient salt has been added. The grain can be better observed, when a sample is taken on a trowel and is allowed to run off, by tilting it at an angle. Granular specks of soap will be seen, intermingled with a watery liquid, which is somewhat caustic to the taste.

An even better test is to place a drop of phenolphthalein on the soap. It will be observed that the watery liquid (spent lye) immediately turns red, but the soap itself, will be much slower in turning pink or reddish. This indicates that the brine is washing the excess caustic soda from the soap. The boiling is continued for another half-day, to ensure perfect saponification of the fats. Then the steam is shut off and the batch is allowed to settle over night.

It is a good practice to fit the open coil with a half-inch outlet at hand level, near the steam valves. While the steam is being used, this outlet is kept closed by means of a pet-cock or small valve. When the boiling is stopped, the pet-cock or valve is opened so as to allow the seething lye to escape from the bottom of the kettle and then recede to its proper level as the boiling stops. By this means, the soap is prevented from sucking into the open steam coil when a vacuum forms in cooling. In other words, the coil remains full of liquid instead of being clogged by the sucked-in hard soap.

On the following morning, the lye is drawn off by the valve at the bottom. This spent lye is rich in glycerol and may be saved for glycerol recovery or, if that is impracticable, it may be sold to a large soap factory. Lye sent out for glycerol recovery should not be too strong. Otherwise, too much sodium sulfate will be introduced into the evaporator when the lye is neutralized, even if we disregard the waste of caustic soda and acid.

Water is again added to almost the level at which the soap stood before withdrawing the lye. The charge is now boiled with the open steam coil and about 50 pounds of 37°Bé lye is added. After boiling through, if a test shows that there is a slight excess of lye present, salt is immediately added to grain the soap once more, and the boiling is continued for several hours. The soap is then allowed to settle as before. This is known as the *strengthening change.*

It is to be noted that when the water is added on the second day, the mass thickens and may become lumpy. To remedy this, salt is added and the boiling continued while adding 10–25 pounds of lye (this is part of the 50 pounds mentioned previously). For economy, the second spent lye, which usually contains more free alkali than the first spent lye, may be used in a subsequent saponification.

After the second or strengthening lye has been drawn off, water is again added to bring the soap to its former level and the process is repeated, adding a little lye and a little salt, boiling for several hours and allowing the soap to settle. By this time, the fats should be completely saponified and nearly all of the coloring matter and impurities washed out. A rapid test for completeness of saponification is made by dissolving about 3 grams of soap in hot distilled water. If the saponification is complete, the solution will be clear, not turbid.

On the fourth day, after the lye is run off, water is added in somewhat smaller volume than on previous occasions and the boiling is continued with open steam coil to give good agitation to the mass of soap. This process is known as *pitching* or *fitting.* When the boiling has proceeded to a point where the soap appears smooth and has a shiny surface, a sample taken on the trowel will run off in a smooth sheet and hang on the end of the trowel in a little pigtail, which should be translucent.

The boiling is continued for several hours, after which the soap is allowed to settle for a day or even longer, depending on the atmospheric temperature (for cooling down to 180–190°F).

The soap settles in two layers. The top or major portion—which constitutes about 80% of the batch and consists of 67% of anhydrous soap with approximately 33% of water and such minor impurities as sodium chloride and sodium carbonate—is known as *neat soap*. The lower layer, called *nigre*, as the name implies, is darker than the neat soap, and consists of a mixture of about 50% of soap and the balance caustic soda, salt water, and dirt.

The neat soap may be pumped off through the draw-off pipe, which is placed above the nigre layer to draw neat soap only. Thus it is pumped to the crutcher (Fig. 7:2), where air is churned in, to make the soap buoyant. After 10 minutes' stirring, a sample is withdrawn, cooled and immersed in water to test its buoyancy. If the soap floats, perfume is mixed in (3 pounds to a frame) and the soap is drawn off into frames.

Floating Soap

In the manufacture of floating soap, the crutcher is filled to a point under the top of the screw blade. By exposing part of this blade, air is beaten in. Consequently, a frame of floating soap will contain about 1000 pounds of soap, whereas a crutcherful of laundry soap weighs 1200 pounds. To speed aeration, some producers use special crutchers for floating soap.

In about a week, the soap will be hard enough to permit *stripping* the sides of the frames. The soap is then slabbed, cut, and dried in steam-heated tunnel driers to give it a superficial skin. After cooling to room temperature, the soap is stamped and wrapped.

This soap, as drawn from the kettle, contains 33% of moisture. The soapmaker figures that a kettle yield of 150%, calculated on the stock, should result from the saponification. That is from a 4000-pound charge of tallow and coconut oil, the yield should be 6000 pounds of soap. Subtracting 20% for nigre, which remains in the kettle, we have 4800 pounds of pure soap, containing 33% of water. The nigre is used as a basis for the succeeding batch, until it becomes so dirty that it cannot be used for the same grade of soap. It is then washed and transferred to the kettle for laundry soap.

On the whole, floating soaps may be said to be pure; for example, one manufacturer states that his soap is "99.44% pure." Some soapmakers add 10 pounds of liquid silicate to a frame of floating soap. This is added for imparting a smooth lustrous finish to the bar, and also because silicate acts to inhibit yellowing, which is an indication of oxidation and occasional rancidity.

As perfume, a mixture of spike lavender and citronella with a dash of rosemary oil may be used. Modern American taste points toward floral-bouquet odors. All perfume houses sell special blends made to withstand the hot soap without decomposition. One manufacturer states that he adds no perfume to this type of soap, because it is also used for dishwashing, where perfume is objectionable.

Despite the fact that the soap is quite neutral, wrappings should always be alkali-resistant to prevent discoloration of the soap and wrapper. To avoid formation of mold, the paper should be free from starchy matter. Waxed paper wrappings with tight seal prevent warping of the soap and preserves the perfume.

Some years ago, nearly all brands of floating soap contained 25% or more of moisture, so that they warped badly on drying. This was remedied to some extent by sealing the soap in waxed wrappers. Two patents, granted to Lever in 1941, cover apparatus and process for producing floating soap, considerably lower in moisture content.[1] In one patent, milled particles of soap are enclosed under pressure, while air is beaten in. The soap is then extruded, causing buoyancy in the bar when the pressure is released. In the other patent, the air is forced into the plastic soap while it is under pressure.

Laundry Soap in Bars

Yellow Laundry Soap (P-S 591*e*) has lost favor with both the manufacturer and the householder. We might discuss it, however, from the standpoint of the Federal specification.

Yellow laundry soap, otherwise known to the trade as *rosin soap,* is usually produced by the boiled process. The stock consists of tallow, house grease, or even certain fatty acids of fairly high titer (40°C or over), and rosin. The fatty portion of high grade rosin soap may consist of 65% of fat and 35% of rosin. The finished soap, therefore, will contain about 25% of rosin, and will conform to the Federal specifications.

The choice of fats and the particular type of rosin remains with the manufacturer. If a light-colored soap is to be made, tallow and a light grade of rosin will be used. Such materials will be costlier than grease and K rosin, which produce a darker product.

The fat is killed in a manner similar to that described for floating soap. From one to three changes are made, depending on the stock used and on whether it is advisable to extract glycerol. After withdrawing the spent lye:

1) The rosin may be saponified in a separate kettle by means of sal soda (concentrated solution of soda ash) and then pumped to the kettle containing the saponified tallow; or

2) The rosin may be broken up, shoveled directly into the kettle of soap, and then saponified with lye.

Rosin requires the same amount of alkali for saponification as does tallow. The lye is added in small increments, as previously suggested, until all the rosin has been saponified as indicated by tests previously described. It is to be noted that all fatty acids may be saponified with sal soda. Rosin is chiefly abietic acid, but there is considerable foaming, because of the evolution of carbon dioxide (CO_2) when sal soda is used. Therefore, sufficient space should be allowed in the kettle, so that it does not boil over. Because of this possibility, many soapmakers forego the economy of saponifying with sal soda, and use caustic soda. In addition, it is doubtful that there would be any economy except in large-scale production.

The soap is salted out as usual, and at least two more washes of a few hours each (for a small kettle), are given. The soap is then ready to be "fitted," or to be given the settling change or *pitch*, as it is sometimes called. Water is now added, and the soap is boiled by means of the closed coils, with only a slight "crack" of the open coil valve, for agitation. The soap should boil smoothly and with a nice sheen on the surface. The amount of water required can be ascertained only by experience.

As noted previously, the settled soap must contain 33% of moisture. When the last lye has been drawn off, however, there is already some water in the soap; that is, the soap that remains in the kettle is not anhydrous. Therefore about 10% of water, on the basis of stock saponified, may be added safely and the mass brought to a boil. If it boils smoothly, and the soap appears to be of the right texture—sliding off the trowel and leaving a pigtail—there is sufficient water. At this point, a rapid test for moisture may be made by the Stark and Dean distillation method, if desired (see Chapter 6). A sample taken before the nigre has settled will have 1–2% more moisture. After settling for one or two hours, a sample may be taken out in a small box, to see how the soap sets up. It should be remembered, however, that the finished soap with the filler will be considerably harder than the basic soap.

The soap is allowed to settle over night and if the temperature has fallen to about 180°F, it may be pumped to the crutcher (see P-S 591*e*, *Composition of Crutcher Batch*).

Soda ash is added little by little, while the screw is being turned, until all the soda ash is absorbed and the soap is uniform. N Brand silicate is added next and mixed until all is uniform. Finally mineral oil and perfume are added and crutched in. When the soap becomes smooth, it is withdrawn to a frame.

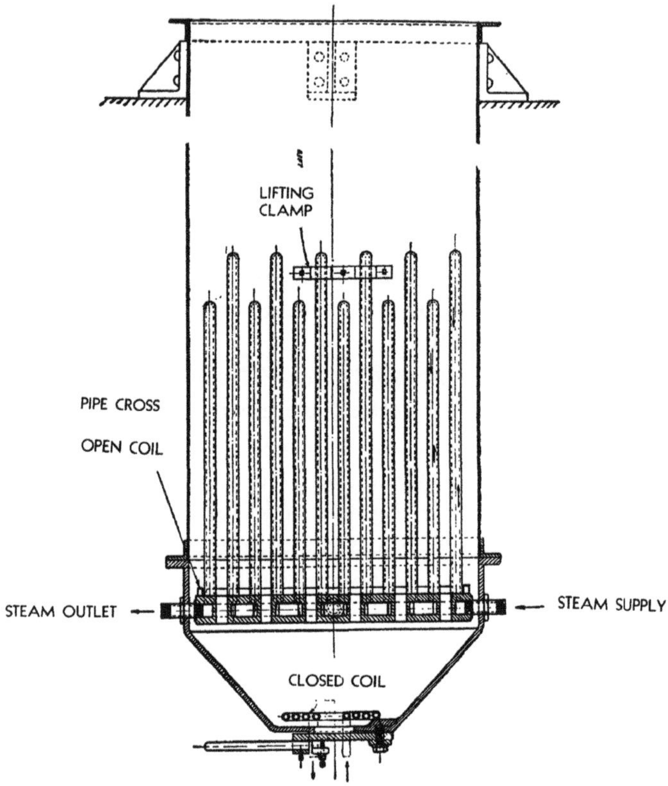

FIG. 7:9. REMELTER FOR SOAP SCRAP (Houchin)
(*Benjamin Eastwood Co., Paterson, N.J.*)

The mineral oil used is yellow and somewhat thinner than petrolatum. This addition imparts a smooth glossy finish to the soap. Possibly it reduces the stickiness of rosin soap.

Rosin soap should be allowed to stand in the frame for at least a week, until it is thoroughly solidified throughout, before stripping the frames. After cutting, the bars should be racked off and placed in a tunnel dryer over night, cooled, and then pressed. In pressing, it is common practice to allow brine to drip on the dies of a power press, or to wet one out of six bars with brine, in foot pressing. This prevents the soap from sticking to the die.

Soap Scrap

Soap scrap, consisting of imperfect bars, ends, and spilled soap, may be handled in several ways. If the scrap is clean, it may be treated in the *remelter* (Fig. 7:9), which is a tank with a conical bottom, having numerous upright steam pipes; or the scrap may be charged into the crutcher, where it is heated and crutched slowly until it is entirely remelted and uniform in consistency, so that it may again be framed.

If the scrap is dirty, it may be thrown back into the kettle and reboiled. In this case, the filler, if any, will be wasted.

Rapid Saponification

The process of making boiled soaps is rather slow and, although many patents have been granted for speeding saponification much of the world's supply of soap is still produced by the method previously described. Automation, however, is rapidly gaining headway.

Many of the rapid and continuous methods employ high temperatures and pressure. In many cases, fatty acids produced by high-pressure hydrolysis of oils and fats are used as soap stock. Patents granted to Victor Mills and Procter & Gamble Co.,[2] illustrate continuous processes for high-pressure fat-splitting and saponification.

In these methods, the fat and water are metered and pumped to heaters operating at a temperature of about 500°F. The pumps force the oil and water into a long reaction chamber at a pressure of 150–1600 psi. A suitable catalyst, such as zinc oxide, is added. The fat enters the top of the autoclave as the steam enters from below. The hydrolysis takes place as the water and fat move countercurrently in the chamber which is 2.5 feet wide and 50 feet high. The reaction time is about an hour, for splitting 7500 pounds of fat. Sweet water, containing 24% of glycerol, is recovered from the bottom of the tower. The fatty acids are drawn off at the top.

The hot fatty acids may now be saponified by bringing them into contact with a measured supply of caustic soda. The hot soap is then sprayed into a tower, where it cools and is converted to granules of soap.

Patents of Clayton and Thurman[3] cover continuous saponification of neutral fats and the recovery of glycerol by volatilization. The reaction time is only five minutes, using fat at 275°F and strong lye at

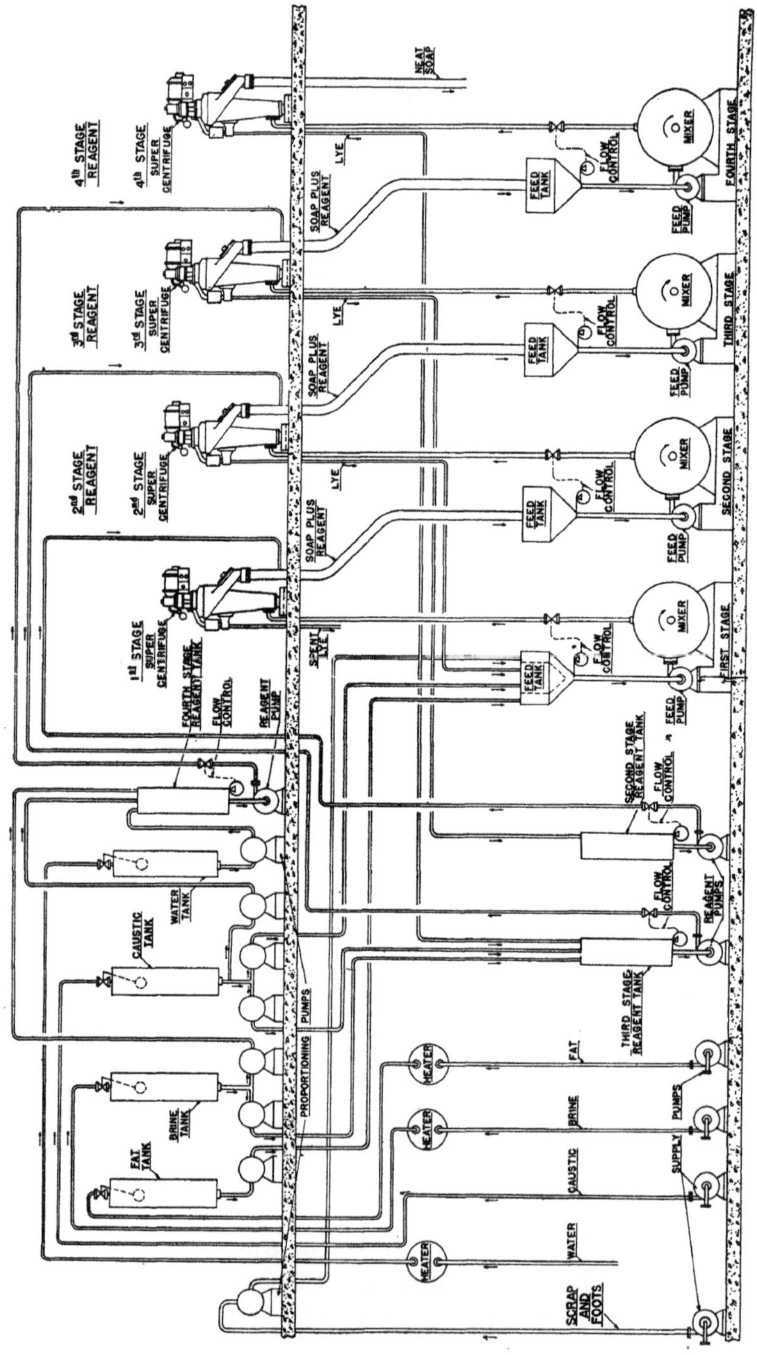

FIG. 7:10. CENTRIFUGAL PROCESS FOR MAKING SOAP
(Sharples Div., Pennsalt Chemical Corp., Philadelphia, Pa.)

180°F, under high pressure. The dry soap is discharged at 550°F into a vacuum chamber, where glycerol and water are flashed off, leaving practically anhydrous soap. This soap is rehydrated, cooled to 200°F, and may then be extruded through a plodder for milling.

The Sharples Specialty Company's patents cover saponification in an autoclave but at relatively low pressure. The reaction takes place in a special mixer into which the fat and lye are entered in the presence of brine. A 95% conversion is effected in one-half hour. The soap and brine are centrifuged. Saponification is completed in another mixer. The soap is washed twice more with brine, and it is finally finished on a nigre. The nigre is separated from the neat soap by a fourth centrifuge. The entire process requires only a few hours. It yields a lye with a fair concentration of glycerol.

Nine sizes of plants are available from Sharples. These plants process from 1500 to 13,500 pounds of fat an hour. The smallest plant, shown in Figure 7:10 requires a space of 25 x 35 x 33 feet.

PROCESS FOR HALF-BOILED SOAP

In preparing half-boiled soap, all the ingredients are charged into the kettle and saponified by heating. The process may be carried out in the regular kettle, a jacketed kettle or a crutcher.

In this process generally, no glycerol is recovered, even if glycerides are saponified. Whatever ingredients are used remain in the finished soap. There are, however, some processes that include rapid saponification with recovery of glycerol; for example, the Descha process of Julius Schaal features the graining out of the saponified mass with salt and lye. The soap contains less water than the full-boiled soap and may be milled to toilet soap without further drying. The color of the resulting soap is not quite so good as that made by the whole-boiled process.

Although the half-boiled method is used for producing all soft (potash) soaps, hard soap may also be so produced. Much laundry soap used to be made by this process.

The fats or oils with the necessary amount of water are charged into the kettle or crutcher and heated to about 150°F. The weighed amount of lye is run in slowly, while the whole mass is agitated with steam in a large kettle or by means of the screw in the crutcher. The constant agitation prevents the formation of "water-balls"—a mixture of hydrated soap and fat—(sometimes called "bunching"), and also prevents sudden rise of the soap in the kettle. When the boiling has continued for some time and the charge appears to be uniform and has the consistency of settled soap, a sample is taken out and tested for completeness of saponification, as previously indicated under whole-boiled soap, Where soap is required to be neutral as in soft soap, it is tested for free caustic soda or potash, and if necessary, adjustment should be made.

Potash Soap

A typical example is P-S-560: *Soap, Automobile and Floor, Liquid and Paste* (*q.v.*). The outline for manufacture is given under this specification, elsewhere in the book.

SOAPMAKING BY COLD PROCESS

The cold process of soapmaking is similar to the half-boiled process, in that all of the ingredients that enter the crutcher are withdrawn with the soap. There is no recovery of glycerol.

The cold process differs from the others in that fatty acids cannot be used. Although stock either high or low in free fatty acids may be used in the whole and half-boiled processes, the stock used in cold-made soap must not exceed 6 or 7% of free fatty acids. An excess would cause the lye to combine with the fatty acids so rapidly, that the mass would bunch and become lumpy. In this process, no heat is applied in the crutcher and the soap is always made in small batches that is, a crutcherful at a time.

If several frames are to be made in one run, the stock is melted in a separate tank and the requisite amount is pumped to the crutcher. For example, laundry bar soap would be made by the cold process as follows:

The fats consist of 15% coconut oil and 85% of tallow. These are heated to 130°F under moderate climatic conditions, and to 120°F in a warm climate. Pump 575 pounds of the mixed fats into the crutcher and start stirring. Add at once 375 pounds of 32-Bé caustic soda and 250 pounds of sodium silicate. Stir for two minutes, or until the mixture is quite well emulsified and looks like a heavy cream. Add 2 pounds of suitable perfume and run the mass immediately into a well-stuffed frame.

The frame should be previously luted or stuffed with mashed-up rosin soap, which has the consistency of putty. The soap will be fluid for several hours, so that unless the frames are made leakproof, there is considerable loss. There will be a steady rise of temperature in the frame for two or three hours, during which the soap may rise in a small hump at the top of the frame. As it cools, this will recede. After two days, the frames may be stripped. The soap may then be allowed to cool for another day, when it will be ready for slabbing and cutting. As cut, the soap will have the following approximate composition:

Moisture	36.75%
Sodium silicate	7.50
Excess Na_2O	0.75
Anhydrous soap	50.00
Glycerol	5.00
	100.00%

Toilet soap may also be made by this process, although as far as large-scale production is concerned, cold-made soap is a thing of the past.

CONTINUOUS PROCESS OF SOAPMAKING

The standard process of making soap in the kettle, has been described at length. For this aerojet age, however, we now have some completely automated soap plants, wherein the same amount of soap is produced in two hours, that formerly required two to five days by traditional methods.

Joseph Crossfield & Sons, Ltd., at Warrington, England, have been successfully operating their new plant, which yields 645,000 pounds a day, of soap base from nearly twenty materials. This plant, which is said to be (1962) the largest automated soap-producing facility in the world, is run by six persons. Its features are: The control system of the blending and saponification sections; widespread use of centrifuges for treatment of spent lye; separation of soap from nigres and brine, and the recovery of glycerol.

Figure 7:11 shows the equipment of the Mazzoni installation, which is made by G. Mazzoni S.p.A., Busto Arsizio (Varese), Italy, for the continuous production of toilet soap.

SOAP PRODUCTS

Soap Powder

Because of its manifold uses and consequently its wide sale, soap powder is produced by almost every manufacturer of hard soap. It must be admitted, that as far as household use is concerned, it is almost unknown to the present generation of householders, having been displaced by synthetic detergents. The United States Government bought vast quantities of this type of product which is covered by specification P-S-606a. It is an inexpensive powder which is economical for cleansing very soiled surfaces or fabrics. It contains not less than 15% of anhydrous soap, and the sum of the alkaline salts calculated as sodium carbonate and anhydrous soap is not less than 55 percent. The remainder is usually moisture.

As the refining of cottonseed oil developed rapidly after 1850, it

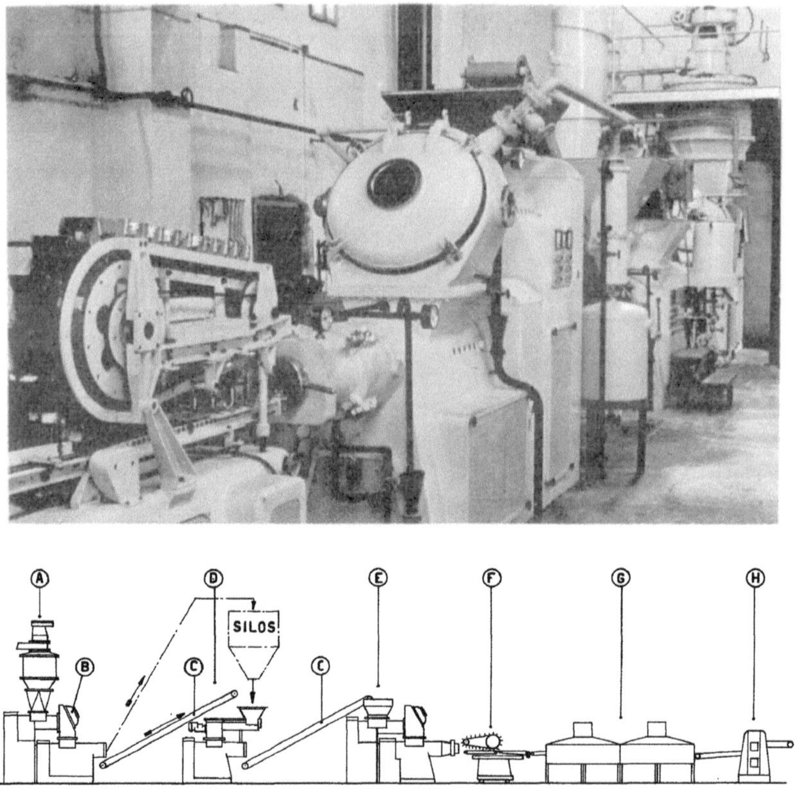

FIG. 7:11. PLANT FOR CONTINUOUS PRODUCTION OF TOILET SOAP
Upper: Shows conditioning tunnel, cutter, duplex plodder, conveyor, continuous amalgamator, conveyor, vacuum spray-dryer
Lower: *a*, Vacuum drying; *b*, Preliminary extrusion and refining; *c*, Continuous belt conveyors; *d*, Continuous mixer (MC); *e*, Final extrusion and refining; *f*, Continuous automatic cutter (TV); *g*, Cooling tunnel (TR); *h*, Automatic stamper
(*G. Mazzoni, S.p.A., Busto-Arsizio, Italy*)

was found that the *foots*, which is a by-product, could serve as the fatty material for the manufacture of soap powder.

Whether James Pyle, who is credited as the originator of soap powder in the United States, actually used cottonseed foots, is not definitely known. He may have employed other fatty materials, as is done when there is a shortage of foots. As foots is a very cheap soap stock, it was admirably suited for this purpose.

Method of Manufacture

Cottonseed foots is a soft, pasty yellow to brown mass, having the appearance of prepared mustard. It consists of about 50% of total fats, partly saponified and partly in the form of occluded neutral oil, with considerable water and organic impurities from the seeds. This material is charged into the kettle and treated as for making a whole boiled soap. As the saponification has converted all the free oil into soap, and as most of the impurities have been washed out in the process, the settled soap is ready for conversion into soap powder.

The process depends on the fact that soda ash absorbs almost twice its weight of water to form *sal soda* crystals. In other words, the anhydrous soda ash (Na_2CO_3) which has a molecular weight of 106, is converted to its crystalline form ($Na_2CO_3 \cdot 10\ H_2O$), whose molecular weight is 286. Obviously, 106 pounds of soda ash absorbs 180 pounds of water.

To make a 1200-pound batch of soap powder, 270 pounds of boiled soap is pumped into a crutcher, while stirring; 480 pounds of soda ash is shovelled in, and at the same time 450 pounds of water is added. Cold water may be used, but the mass is usually kept plastic by turning on the steam. Heat is evolved while the soda ash is going into solution. When the mass is uniform, it may be handled in several ways. It may be run into frames and there allowed to crystallize while cooling. This method of manufacture will not require great floor space. More time is consumed in cooling, however, than when the mass is poured out on a concrete floor. The second method has its drawback in that great floor space is required.

If the mass is allowed to cool in frames, the powder must be cut into blocks before it becomes too hard. It is cut manually by means of wires. The blocks are stacked, allowed to cool, and in a few days, they begin to crumble. At this point the mass can be easily broken into size ready for entry into the preliminary crusher, from which it is conveyed into the grinder.

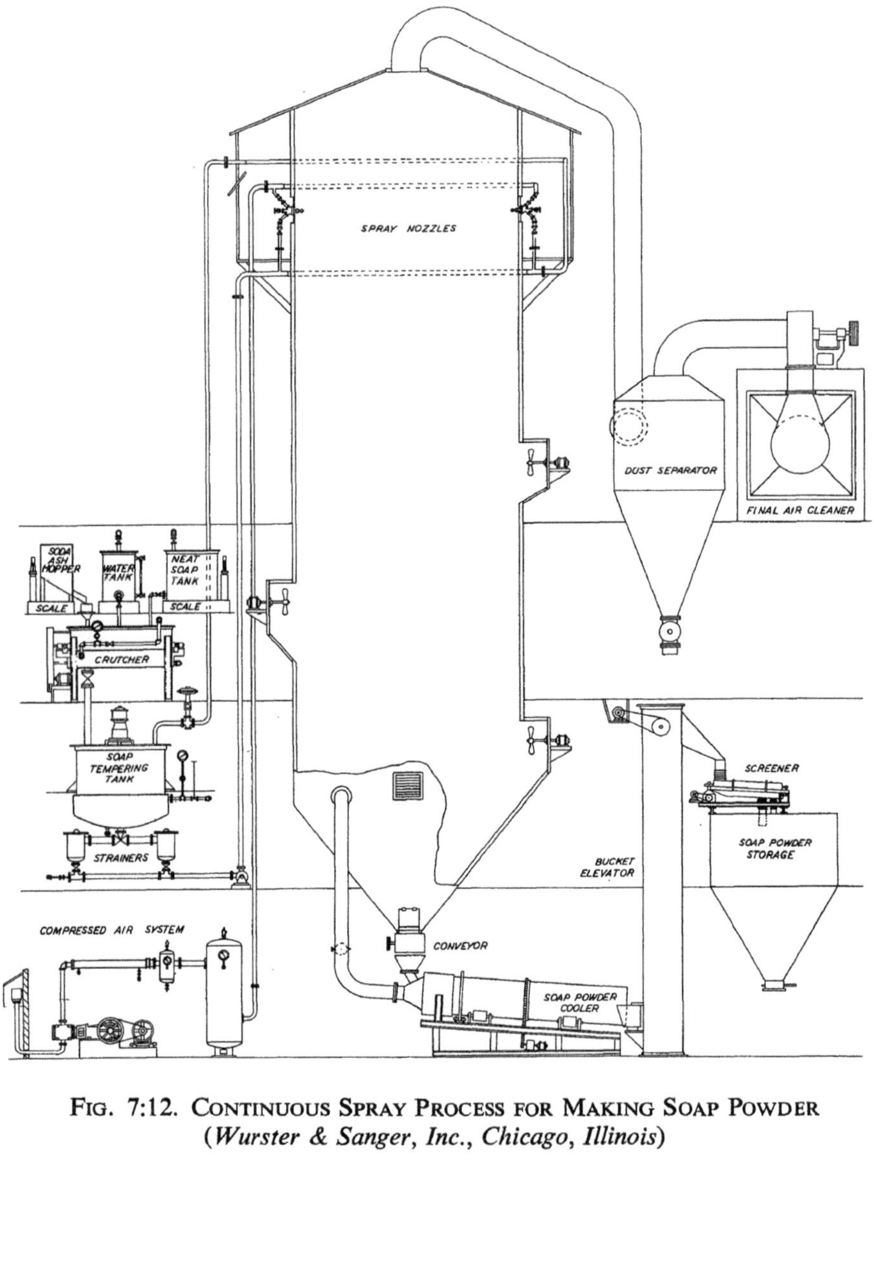

FIG. 7:12. CONTINUOUS SPRAY PROCESS FOR MAKING SOAP POWDER
(*Wurster & Sanger, Inc., Chicago, Illinois*)

If crystallization takes place on the floor, the material must be turned over several times before it hardens sufficiently to grind.

To avoid so much manual labor, the next step in the development of manufacturing soap powder was to run the hot mix on a brine-chilled rotating steel roller. The powder, which crystallizes in a thin film, is scraped off by means of a knife. The flakes of powder are conveyed into a hammer-mill.

The chilling roll was superseded by the spray-crystallizing, or spray-chilling, process.

Before spraying a product of this sort, it must be heated and thinned out considerably so as to pass through the spray nozzles. Many devices for accomplishing this are covered by patents. Usually a blast of cold air meets the spray as it descends. In the whirl, the particles crystallize and drop to the bottom of a tower and may be packaged immediately.

Powder produced by all previous methods was rather dense and therefore had small bulk. The powder produced by spraying, consists of small spherical spongy particles, having considerably more bulk than that produced by former methods. To produce a powder with high moisture content, the spraying chamber is cooled by mechanical refrigeration. The moisture content of the finished product can be be regulated to 45% or lower, as desired. [4]

Bulk is particularly valuable for retail sales. Sprayed products have been well accepted by consumers.

Although soap powders containing a high content of anhydrous soap have disappeared from the retail market, having for the most part been displaced by synthetic detergents, the spray dryer can also produce both filled and pure soaps (like Ivory Snow).

Figure 7:12 shows an installation for the spray-drying of soap powder.

Soap Flakes and Powder for Industrial Use

Soap flakes and soap granules have lost much of their favor with the modern housewife. In the household, most of this type of soap has been supplanted by either light- or heavy-duty synthetic detergents. Consequently, soap flakes and ground powders are used in power laundries, in textile mills, and in industry—for example, for the burnishing and drawing of metals.

P-S-580 is a prototype of the chips that are used in power laundries. Although the minimum titer allowable is 39°C, this soap is usually made with tallow. The composition of one batch for a crutcher is given under the specification in the latter part of this chapter.

In small installations, flakes may be produced by framing the soap, then slabbing and cutting into long bars, which are then stacked, for preliminary drying. The bars are then chipped on a chipping wheel

FIG. 7:13. SOAP CHIPPER (Houchin)
Left: Six blades; right: twelve blades
(*Benjamin Eastwood Co., Paterson, N.J.*)

(Fig. 7:13), and the chips may be dried on screens set into a cabinet dryer. This process is archaic and requires much labor. Circumstances may warrant such procedure. Also, the chips are much thicker than those produced on a Proctor or similar dryer. On the other hand, the bars may be run through a plodder to produce thin flakes, which may then be dried to a low moisture content. The flakes may be ground to a powder, if required.

SPECIAL SOAPS

Milled Toilet Soaps

Most toilet soaps are made by what is known as the *French Milled* process. By the conventional method, a boiled soap, somewhat similar in composition to floating soap, but usually containing less coconut oil, is used as the base.

The settled soap is dried as flakes, in a Proctor or similar dryer, to a moisture content of 12 percent. These chips or flakes are generally stored for a few days in a large bin, so as to yield soap of uniform moisture content. A batch of 200 pounds of flakes, is placed in the amalgamator (Fig. 7:14), which is a horizontal trough-like mixer, from which the processed soap, perfume, color, and any other additive may be dumped. An aqueous dye solution which has been strained through muslin and 1% of a floral bouquet perfume oil, 1 pound of pure titanium dioxide and one-half ounce of Soapanox, are added while the mixer is going. In a few minutes the entire charge is uniformly mixed.

FIG. 7:14. ASSEMBLY FOR CONTINUOUS MILLING AND PLODDING (Houchin) (*Benjamin Eastwood Co., Paterson, N.J.*)

The flakes are now run into the milling machine, the rolls of which may be either of stone or steel (Fig. 7:14). The soap, in long ribbons, usually runs from one set of rolls to another until the color and perfume have been uniformly distributed. These ribbons are then run through a plodder, which is a heavy machine somewhat on the order of a meat grinder, which extrudes the soap in long bars. Each bar is placed on a cutting table and cut into small 3 to 4 oz. cakes, which may then be pressed to the desired shape and wrapped. Figure 7:15

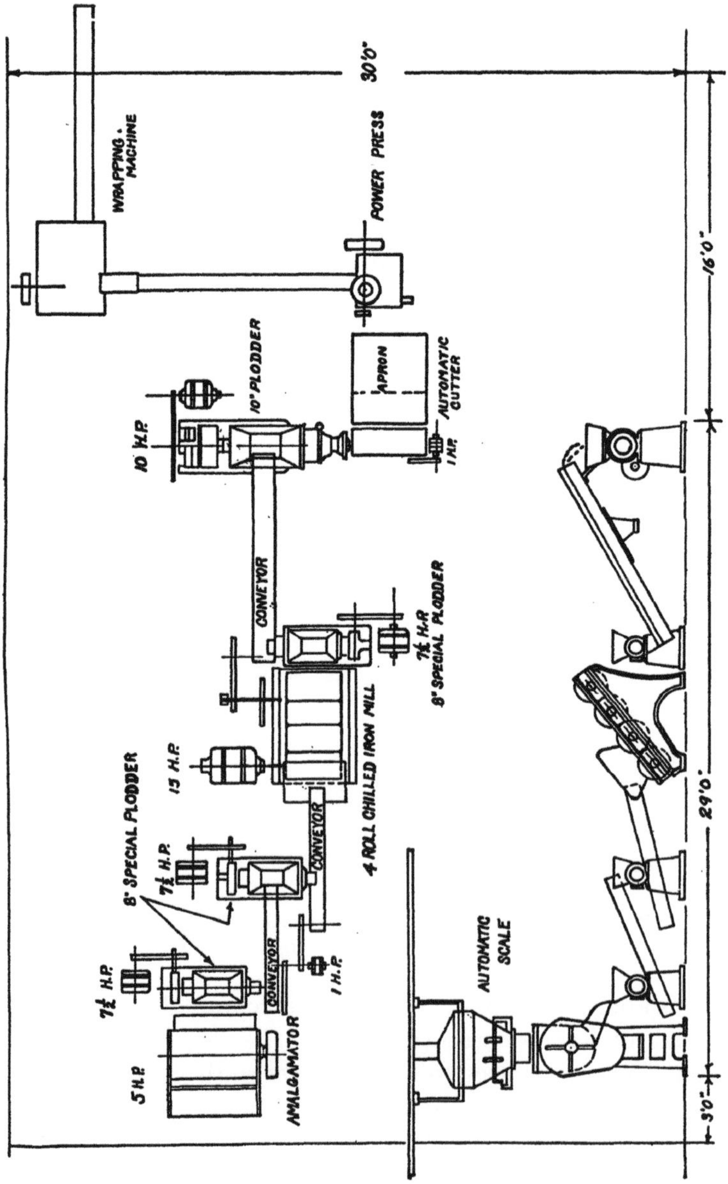

FIG. 7:15. DETAILS OF LAYOUT FOR MAKING MILLED TOILET SOAP (Houchin)
(Benjamin Eastwood Co., Paterson, N.J.)

shows the layout of equipment for the manufacture of milled toilet soap, which includes two special plodders between the amalgamator and mill (Fig. 7:16).

Milled soaps may be made with special ingredients such as oatmeal, sulfur, superfatting agents—lanolin and its derivatives—pine tar, cresylic acid, bacteriostats such as hexachlorophene, Actamer, Otto-sept, etc. In each case, the special ingredient is added in the amalgamator and the flakes are then milled as described.

FIG. 7:16. SOAP PLODDER (Houchin)
(*Benjamin Eastwood Co., Paterson, N.J.*)

The special properties ascribed to this type of soap, other than properties due to active ingredients such as those mentioned, are due, firstly, to the fact that in being worked through the various machines, any traces of free caustic alkali are carbonated by the atmosphere. This tends to reduce its alkaline strength. Secondly, a greater amount of perfume can be admixed with the cold soap, than is possible in the crutcher of hot soap and less of the volatile ingredients is lost. A more uniform distribution of solid ingredients is possible by milling. Finally, as the soap contains less water it does not warp and wash away so quickly as floating soap.

A fault occasionally found in milled soap, is that it cracks as it washes down to a small thin piece. These cracks or striations fill with soil so that the cake has an unsightly appearance. These cracks have been attributed to occluded air, between the ribbons of soap being plodded. Differences in moisture content of the soap ribbons may also cause uneven cohesion.

British patent 512 551, assigned to Colgate-Palmolive Co., is said to overcome this difficulty. In this patent and in USP 2 213 772, assigned to Procter & Gamble, the soap is plodded under reduced pressure to yield a product free from striations.

Castile Soap

Although the composition of castile soap has changed, since its origin in Castile, Spain, and during the years of litigation between the Federal Trade Commission and the soap manufacturers, as to what should be considered as Castile soap, nevertheless, the public still thinks of Castile soap as a product made entirely of olive oil. Moreover, it is the impression of this author that as the consumer pays the bill, soap made entirely of olive oil or green olive oil, should be designated as *pure white*, or *pure green, castile soap*, respectively. Since the Federal Trade Commission ruled (1926) that any nice white soap that is pure may be designated as castile soap, the consumer often acquires soaps consisting of coconut oil with other oils and fats, when he buys "Castile Soap."

One authority, F. E. Wall,[5] says: "Soaps made with olive oil are generally conceded to be best for the skin." No argument is presented here as to whether or not castile soap, according to the author's definition, is better than any other soap. It is true that olive oil soap washes away quickly, because of its low titer, and does not produce a profuse lather. Most people, however, find that olive oil soap is mild and perhaps gentler to their skin than other types of toilet soap.

For this reason, hairdressers also have preferred olive oil soap, but for shampooing also, olive oil soap has been largely displaced by the synthetic detergents.

Hairdressers' Shampoo

What the hairdressing industry calls *coconut base soap* is a soap made of coconut oil and potash only. When finished, it contains 45% of soap and glycerol and 55% of water. Stated otherwise, a pound of coconut oil yields 2.5 pounds of this base.

A characteristic of this soap is that, while boiling it is quite thin and at the end of the saponification, it runs freely from the kettle, thus lending itself easily to filling in 5-pound cans, where it solidifies on cooling. If such a can of congealed soap is tapped with the hand, there is a distinct quiver and a drum-like sound. No other soap (with the possible exception of waterless hand soap) has this character. Furthermore, the soap is almost transparent. Here is how it is made:

Ceylon or crude coconut oil	40 lb
Water	37
Caustic potash (45%)	23
	100 lb

The procedure is about the same as above. When finished, it may be colored or perfumed as required. The soap test should be for neutrality and moisture before it is run from the kettle. If water-white soap is required, cochin type oil should be used.

As labor costs have increased, fewer barbers or hairdressers are making their own shampoos, so this particular type of soap base is no longer of as great interest as formerly.

The soap may be used as a base for liquid hand soap in dispensers, and as a scrub soap. To facilitate the latter purpose, the soap is run into an open-head drum, which is equipped with a ¾-inch bung on the side and near the bottom, and a 2-inch bung in the removable head.

A bar of wood 4 × 4 inches in cross section and 36 inches long, with a 1½-inch hole bored through it about 3 inches from one end, is set, hole up, in the drum, directly in front of the small bung, before the soap has congealed. It must be so fixed that it does not move out of place. The hole in the bar is necessary for putting through a crow-bar or rope with which to pull out the bar, when the soap has congealed, leaving a channel all the way down the side of the drum. (It may be necessary to use a block and fall to remove the bar.) The drum is headed when the soap is cold. The small bung near the bottom should be kept clear of soap. At the point of use, a spigot is inserted in place of the small bung. Water is run into the top and channel, to produce liquid soap. Each time liquid soap is withdrawn, water is again added, so that there is a constant supply of liquid soap, for janitorial or other use.

A stiff vegetable oil soap (not coconut oil) which can be produced and distributed in "channel" or "system" drums as described, is compounded as follows:

Soybean or corn oil	50 lb
Caustic soda (50%)	2.8
Potash (45%)	17.4
Water	29.8
	100.0 lb

Some tall oil may be included in such a formulation, but we recommend that this be tried, to determine the stability of the channel over an extended period.

Mottled and Variegated Soaps

Mottled soaps have not come into great vogue in the United States. With the influx of Latin Americans, however, some mottled soaps like those to which they were accustomed in their home lands, have been produced here to satisfy the demand. Mottled soaps have been used in Europe for many years.

The original Eschweger soap was a whole-boiled soap, made with coconut oil and tallow. This was thoroughly saponified and left on a coarse grain, slightly alkaline. About four ounces of ultramarine blue was mixed with a half-gallon of water, containing 0.1% of caustic soda, based on weight of soap, for each 1000 pound. This was crutched into the hot soap, in great boxes containing several thousand pounds. The batch was left to harden and mottle, and it was then cut into bars.

Later methods improved this process so that from one batch of soap, mottles in several colors were produced, one frame at a time.

Where coconut soap is desired, as for kosher soap, the coconut oil is saponified in the crutcher. The design is then produced, mixing vermilion red or ultramarine blue with coconut oil and running the colored oil into the hot soap in the frame by means of a special device to produce the colored trade-mark or design. If a mere mottle is desired, the colored oil is poured over the full frame of hot soap and then crutched in by hand.

In the formula given in the section on soap made by the cold process, the author devised a so-called artificial mottle. About a gallon of the mixed oil and fat is kept out of the formula. After the emulsion has been run into the frame, the gallon of warm oil and tallow is mixed with four ounces of ultramarine blue, poured over the top of the soap emulsion in the frame and quickly stirred in, with perpendicular strokes of the hand crutch. The soap, when cold, will show beautiful variegations which are different from the true mottle in Eschweger soap.

Other mottling agents may also be used; for example, yellow ochre, ferrous sulfate (greenish, and finally reddish-brown on exposure), Venetian red, etc.

Metallic Soap

The common idea of soap, is that of a substance that dissolves in water, producing a lather and as being useful as a detergent. The metallic soaps, however, actually repel water. Some of these soaps are soluble only in certain organic solvents such as oils, aliphatic and aromatic hydrocarbons and their derivatives, turpentine, etc.*

As a rule, the solubility of metallic soaps in water, decreases with the increase in the number of carbon atoms in the molecule. The alkaline earth and heavy-metal salts of the lowest fatty acids are soluble in water, readily forming basic salts which separate out of the solutions on boiling. For example, calcium butyrate, $Ca(CH_3CH_2CH_2COO)_2$ is soluble in 3.5 parts of water at 14°C whereas calcium laurate, $Ca(C_{11}H_{23}COO)_2$, requires about 2000 parts of water at 100°C for complete solution.

The barium, calcium, and magnesium salts of the volatile fatty acids are soluble in water.

The lead salts of the higher fatty acids are almost insoluble in ether, whereas lead oleate, linoleate, and ricinoleate are soluble. This furnishes a means of separating the saturated from the unsaturated fatty acids and it is actually employed in analytical procedures.

The most generally employed method for the production of metallic soaps is by double decomposition of the alkaline soaps with aqueous solutions of metallic salts. They may also be produced by heating the free fatty acids with oxides or carbonates of the metals. The physical appearance of the products of this second process, however, do not lend themselves to easy handling for the varied purposes to which they are put. The usual method for producing metallic soap is illustrated by the following reaction:

$$2\,C_{17}H_{35}COONa + CaCl_2 \rightarrow Ca(C_{17}H_{35}COO)_2 + 2\,NaCl$$

The solubility of metallic soaps in solvents is enhanced by the presence of free fatty acids in solution. They have a great many applications in industry. They are used as driers in paint, varnish, and printing inks; as thickeners of petroleum oils for lubrication; for waterproofing textiles, metal-drawing; for extruding and forming plastics; for dusting molds of rubber to prevent adhesion to the metal; in dusting powders for toilet use; and in cosmetics.

Aluminum soaps are used to increase the viscosity or body of an oil.

* Suppliers: 85, 88, 101

The stearate and palmitate are superior to the oleate and naphthenate in this respect. Thickening agents are used in stains to minimize penetration and to produce deeper color effects. Solutions of aluminum soap are used with thinners to waterproof concrete and stucco. They give body to flat paints without materially increasing the pigment content. They assist in suspending the pigment.

Aluminum 12-hydroxystearate which is more soluble in polar and less soluble in nonpolar solvents than the usual form of stearate, with the exception of methanol, ethyl acetate and Cellosolve. It is used in paints, inks, waterproofing, rope lubrication, cement treatment, plastics, leather, but not for lacquer formulation.

Aluminum resinate is used for sizing paper.

Barium stearate is insoluble in polar and nonpolar liquids. It is used in dry lubrication, molding plastics, greaseless bearings, compounding waxes, wire-drawing, or where a waxy, non-tacky substance is required.

Although calcium-based lubricants have been used for many years, the calcium soaps are being displaced by other metallic soaps for use in extreme-pressure lubricants. Aluminum stearate has largely supplanted lime soaps in lubricants.

Cerium stearate, $Ce(C_{17}H_{35}COO)_3$ is a white powder, melting at 100–110°C, insoluble in polar liquids, partly soluble in benzol and petroleum oil. It is used as a catalyst. Other uses are the same as for barium and calcium stearates.

Chromium oleate is a green, waxy substance, soluble in mineral spirits and oils. It is used in compounding greases and for supplying a soluble form of chromium for impregnating fabrics.

Chromium stearate is a dark green powder, melting at 95 to 100°C. It is slightly soluble in benzene and mineral spirits. It is used as a catalyst, in leather treatment, for colored waxes, greases, coatings, and ceramics.

Cobalt, resinate, linoleate, and *naphthenate* are used as driers for paint, varnish, linoleum, etc. With linseed or Chinawood oil, 0.05% is used.

Naphthenated copper hydroxide is a fine green powder, insoluble in petroleum solvents. It is used for treatment of seeds, as a fungicide, and as an anti-fouling agent in paints for ship bottoms.

Copper naphthenate, applied either in water or in Stoddard solvent, protects fabrics completely from mildew, when 20 to 28g are used to the liter. It has a green color and pungent odor, and it stiffens the fabric. *Copper oleate* is used for desulfurizing gasoline and other sulfur-containing oils.

Copper propionyl acetate is used commercially for protection against mildew. It colors the fabric a gray-green.

Lead resinate, *linoleate*, and *naphthenate*, are used to the extent of 1%, as driers.

Lead naphthenate is now used in considerable amounts for compounding with mineral oil in extreme-pressure lubricants on free wheeling, synchro-mesh, and hypoid-worm gears. It is said that this lead soap does not increase the viscosity of the original oil, and that sulfur may also be added to the mixture, in order to increase the strength of the film.

Lead oleate has been used in lead plasters as early as 1653. It is valuable for its adhesive properties, for which it is used in adhesive tape and certain types of insulation and also as a rust-preventive.

Lithium stearate is a white solid, melting at 217°C. It is used in manufacturing grease, lithographing inks, crayons, cosmetics; and as an oil-soluble emulsifying and dispersing agent.

Magnesium stearate is a white powder, melting at 140°C, insoluble in alcohol, benzene, and petroleum hydrocarbons. It is used in ailments of the skin.

Magnesium resinate, *linoleate*, and *naphthenate* are used as driers, to the extent of 0.05 percent.

Manganese stearate is a pink powder, melting at 100–110°C. It is insoluble in polar liquids, partly soluble in petroleum oils, and forms gels with aromatic and aliphatic hydrocarbons. It is used as a catalyst and in compounding waxes and greases.

Nickel stearate is a green powder, melting at 150–160°C. It is insoluble in water and esters, soluble in aromatic hydrocarbons, and forms gels with mineral spirits and petroleum oils. It is used in lubricants, waxes, waterproofing compounds, coatings and plastics, as a catalyst, in leather treatment, as a detergent in organic liquids, and as a powdered quench in nickel welding.

Strontium stearate is a white powder, melting at 130–140°C. It is insoluble in low-molecular-weight alcohols and esters. It is partly soluble in aromatic and aliphatic hydrocarbons, with which it forms gels. It is used in compounding greases and waxes, and for pyrotechnic flares.

Oxidation of lubricating oils is inhibited by 0.2% of *thallium oleate*.

Basic and neutral *tin naphthenates* are amber-colored, viscous liquids, soluble in mineral spirits and oils. They are used in compounding oils and as a source of tin that is soluble in organic solvents.

Tin oleate and *ricinoleate* were found to have antioxidant properties when used in lubricating oil.

Zinc stearate, melting at 120°C, is insoluble in ethanol, benzene and mineral spirits. Because of its mild antiseptic properties and its power of absorbing body odors, it is useful in compounding face, foot, and toilet powders.

**Zinc undecylenate* (also *copper undecylenate*) is used in a *remedial ointment* for athlete's foot.

Textile Soaps

For Cottons:

Cottons are boiled out in kiers, to remove pectins, oleaginous matter, and various impurities. This is accomplished by means of alkalies, using little or no soap at all.[6]

After cotton is spun, woven, and printed, tallow soap is used to finish the goods. Although many manufacturers of cotton goods use cottonseed soap, in solid form, much labor and time can be saved by the use of chips or granulated soap, which dissolve more rapidly.

The soap is made by the boiled process and then run through the dryer. It is sold in bags as 88–92% anhydrous soap.

There are also built soaps used in many cotton-treating processes. Such soaps contain soda ash, sodium silicate, phosphates, etc. However, the anhydrous soap content is usually agreed on by buyer and seller. The builders are added in the crutcher, before the soap is dried. Granulated soaps are produced by grinding the chips.

For Wool:

Wools are treated more gently than are cottons. That is, the soaps are usually neutral, and the solutions are not so hot except in fulling. The wool industry uses both potash and soda soaps.

As has already been mentioned under wool grease, raw wool is the dirtiest of textile fibers, containing as much as 50% of extraneous matter, which consists of wool grease, suint, and almost any type of soil with which the animal may come in contact, including blood stains. All this soil must be removed before spinning.

The wool scourer gives the wool a preliminary water rinse and then

---

* Supplier: 19

scours by running the fibers continuously through a series of "bowls" or tanks, containing potash soap and a small amount of soda ash.

Olive Fig Soap:

Olive soap has for many years been the standard for scouring wool but it has been superseded by soaps made of low-grade vegetable fatty acids, or by potash tall oil soap.

When sulfur olive oil was used (sometimes called "olive foots") it was compounded as follows:

Olive Foots	440 lb
45% Potash	195
Water	355
Soda ash, or Pearl ash (K_2CO_3)	10
	1000 lb

The procedure is the same as for other potash soap. After saponification is complete, the soda ash or pearl ash is dissolved in hot water and crutched in. This type of soap contains about 0.5% free caustic potash besides the pearl ash, because the wool, being greasy, absorbs or neutralizes the free alkalies. If absolutely neutral soap were used, the fatty matters would acidify the soap so as to make it ineffective.

Low-Titer Soda Soap:

There is considerable fatty oil, such as lard oil and also mineral oil, used in spinning wool. These must be removed before further processing, dyeing, or finishing. Red oil soda soap is used for this purpose.

The soap is produced by the boiled process and run through a dryer. However, low-titer soap does not run easily on the dryer. The soap does not chill on the roll and even when it does run, it is apt to melt on the screens in the dryer. Very cold water run through the roll may have the desired result. The soap may be hardened somewhat by using a small percentage of coconut oil or fatty acids in the formulation. See specification P-S-578 Low-titer Built Laundry Soap, for another formulation.

Fulling Soap:

The *fulling* of woolens is a process of beating and running the goods through a washing machine, in a hot solution of soap, until it has shrunk to proper dimensions. Years ago, fuller's earth was used instead of soap. It was found later, that soap wets and cleans the fabric better. Through long usage, palm-oil soap proved to be very useful for this process. Various price changes, however, and a processing tax of three cents a pound encouraged soap manufacturers and wool processors to find substitutes. Both tallow and house grease have been found to serve equally well. The soap (titer 39°C or above) is made from whatever stock, by the boiled process and is sold as solid, in slack barrels; or it may be produced in flake and powdered form.

For Silk:

For processing silk, hard soap, made by the boiled process with sulfur olive oil was the preferred type. As economic conditions changed, however, and the price of olive oil rose, many substitutes for olive oil have been used.

Silk is a tender fiber consisting of the proteins *sericin* and *fibroin*, which requires processing in tepid water.

For years, raw silk was soaked in an emulsion consisting of green olive soap and olive oil. This was superseded by an emulsion of sulfonated oil, often containing mineral oil. The soaking loosens the fibers that are matted together more or less as the skeins of silk are shipped. The loosened fibers may then be reeled and "thrown" on cones.*

Knitted hosiery and broad silk must then be boiled in a soap solution to free the fabric from silk gum and oil. Here too, green olive soap is used.

Soap similar to that made of sulfur olive oil can be formulated with domestic oils or fatty acids. A low-titer oil—20°C or lower—should be used. Much of this type of soap has been displaced by soaps made of acidulated soap stock, colored green. In substituting such oils, which may be more readily oxidized than olive oil, oxidation inhibitors should be used, for example, diphenyl amine, ortho-biphenyl biguanidine (Monsanto 326), Soapanox (Monsanto), *p*-tert-butylphenol, and *p*-tert-amylphenol.

In many silk plants, so-called "boil-off oils" are used. These are mixtures of potash/red oil soaps, together with alkalies, such as soda ash and sodium silicate, glycols, and surface active agents.

AMMONIA SOAPS, ORGANIC AMINES, AMINE SOAPS

The use of caustic soda and caustic potash as saponifying agents has already been discussed earlier. There are also several other chemicals which combine with fatty acids to form soap. These are ammonia and organic amines.

---

* *Thrown silk* consists of several strands that have been twisted into a single thread and wound on cones for knitting and weaving.

Ammonia soaps are soft. They are used as cleaning compounds in metal polishes, in dry cleaning soaps, and in prespotting, and as emulsifiers. They are not so stable as soda and potash soaps.

Ammonia soaps are formed by reacting ammonium hydroxide with fatty acids. Neutral oils do not saponify with ammonia. The volatility of ammonia is of distinct value in some cases, where it is used to saponify a resin, such as shellac. A case in point is its use in the manufacture of No-rub floor wax, in which the ammonia saponifies and disperses the shellac. After the ammonia dries and then volatilizes, the hardness and luster of the original shellac is restored, imparting water-repellency at the same time.

Amine Soaps

Amine soaps are useful in the preparation of emulsions for cosmetic, textile, leather, metal-working, dry-cleaning, floor waxes and other emulsions for various industrial purposes. Such emulsions are distinguished by their small particle size, ease of preparation, non-corrosiveness, wide flexibility in formulation and stability on long storage. The oleates have the outstanding property of being both water- and oil-soluble and can be incorporated in oil and organic solvents to form "soluble" oils, dry-cleaning soaps and textile scouring and wetting agents. The fully neutralized triethanolamine and mixed isopropanolamine soaps have a *p*H of approximately 8. They also act to lower the alkalinity of soda and potash soaps when combined with them. Emulsions and Detergents, 10th Ed. Union Carbide Chemicals Co. gives the following values of solutions of 2.5 grams of soap in 97.5 grams of water:

	*p*H
Sodium stearate	9.9
Monoethanolamine oleate	9.45
Potassium coconut oil soap	9.4
Morpholine stearate	9.05
Triethanolamine stearate	8.87
Triethanolamine oleate	8.8
Triethanolamine coconut oil soap	8.3

Monoethanolamine,

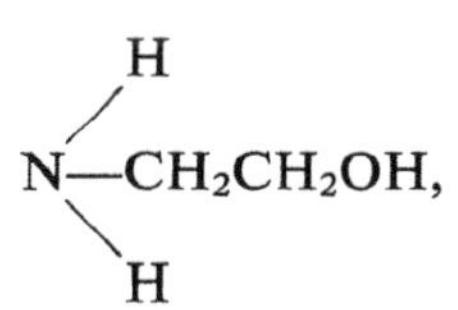

is a slightly viscous liquid, having a distinct ammoniacal odor. It is particularly valuable in those applications where advantage can be taken of its low combining weight. Its molecular weight is 61.06, which means that 61 lb will saponify 282 lb of oleic acid. Monoethanolamine soaps are good detergents, with better lathering properties than triethanolamine soaps. This soap is soluble in hydrocarbons and in mineral oils, which permits its use in making "soluble" oils.

Triethanolamine,

$$N \begin{cases} CH_2CH_2OH \\ —CH_2CH_2OH, \\ CH_2CH_2OH \end{cases}$$

is a viscous liquid with a slightly ammoniacal odor. Besides being completely soluble in water, it is miscible with many organic liquids such as alcohols; it is insoluble in gasoline, benzene, and other hydrocarbons. The average equivalent weight of the commercial mixture (which contains a little mono and diethanolamines) is 142. The pure product as an equivalent or molecular weight of 149. Otherwise, the commercial product differs little from the pure compound. The usual requirements for emulsification are between 2 and 4% triethanolamine and 5 to 15% oleic or stearic acid, each based on the weight of the material to be emulsified.

Mixed Isopropanolamine

Mixed isopropanolamine is a mixture of

Monoisopropanolamine, $NH_2(CH_2CHOHCH_3)$
Diisopropanolamine, $NH(CH_2CHOHCH_3)_2$
Triisopropanolamine, $N(CH_2CHOHCH_3)_3$

The mixture is a viscous hygroscopic liquid with a slightly ammoniacal odor, completely soluble in water but only slightly soluble in hydrocarbons. The combining weight of the mixture is approximately the same as that of commercial triethanolamine, and for this reason it may be substituted weight for weight in most formulas calling for triethanolamine.

Isopropanolamine soaps are more soluble than ethanolamine soaps in hydrocarbons such as mineral oil, naphtha, or gasoline. This increased solubility allows the use of less free fatty acid in "soluble" oils and emulsions. In addition, isopropanolamine soaps have been found to show less darkening in color under normal storage conditions. Their better solubility in hydrocarbons and color-stability are of especial interest in soluble white paraffin oils, where good color and low free fatty acids as well as low *p*H of the oil emulsions are desired.

Morpholine,

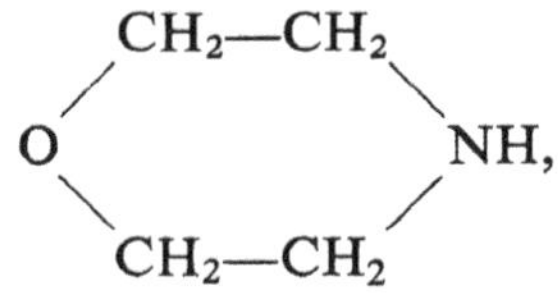

is a colorless, mobile liquid with a penetrating odor that is greatly diminished upon dilution with water. Morpholine forms soaps that are excellent emulsifying agents for use in mixtures where water resistance of the dried emulsion film is desired. The conditions for its use in emulsions resemble those for triethanolamine, and generally a working formula can be obtained by direct substitution of morpholine on an equivalent weight basis for triethanolamine. The equivalent weight of morpholine is 87.

N, N-Dimethyl Ethanolamine,

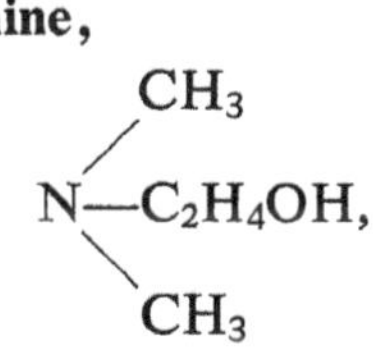

is also a useful component of rubless wax polishes. It is supplied as a 70% aqueous solution for industrial purposes, but an anhydrous grade is also available for pharmaceutical applications. It may be substituted for morpholine, on a contained-weight basis. Wax emulsions made with N,N-dimethyl ethanolamine have the added advantage of not discoloring light-colored linoleum, as sometimes occurs with other amines.

Ethylenediamine,

$$(H_2NCH_2CH_2NH_2),$$

is a colorless water soluble liquid with an ammoniacal odor. It has the lowest combining weight of all the organic amines. Thirty pounds are required to combine with 282 pounds of oleic acid.

In general, it may be said that the amines are valuable for several reasons: ease of preparation, buffered alkalinity, high emulsifying power and solubility in oil. They are useful in the formulation of dry-cleaning soaps, soluble oils, and cosmetic creams.

Diethylene Triamine,

$$(NH_2CH_2CH_2NHCH_2CH_2NH_2),$$

is a hygroscopic, viscous liquid, completely soluble in water and in hydrocarbons. It is strongly alkaline and forms soap with fatty acids. Such soaps dehydrate to amides.

Monoamylene Amine,

$$(C_5H_{11}NH_2),$$

is a colorless liquid having a distinct ammoniacal odor. It is miscible with water and most common solvents. It combines readily with fatty acids to form soaps which are soluble in mineral and vegetable oils. Ninety-six parts of this amine are required to saponify 282 parts of oleic acid. It yields both oil-in-water (o/w) and water-in-oil (w/o) emulsions.

For lubrication of textiles, 3.7% of the oleate soap is thoroughly mixed with a 28°Bé paraffin oil of 100 viscosity, and then stirred with an equal amount of water. For hard waters, amylene amine soaps may be made with coconut fatty acids. A 0.1% solution of the oleate soap has a *p*H of 9.57.

A suggested use for the oleate soap is as a dispersing agent for an insecticide to be used for the control of onion thrips, red spiders, cucumber beetles, tent caterpillars, grape-leaf hoppers, certain aphids, such as are found on nasturtium, green chrysanthemum, golden glow, rose, and green peach. The formulation is as follows:

	%
Monoamylene amine oleate	10
White Oil (80–85 vis)	85
Triethanolamine	5

This makes a clear solution, which is then diluted 1:200 or 1:400 with water. It is said not to injure foliage.

2-Amino-2-methyl-1-propanol, 2-Amino-2-methyl-1-propanediol, and Tri (hydroxymethyl) aminomethane

form soaps that are soluble in water, oils, fats, waxes, and similar substances. They are useful in cosmetics, textile and leather specialties, wax polishes and coatings. The soaps are relatively stable and almost odorless. It is reported that 0.5% solutions of the oleates show a surface tension of 27 dynes to the square centimeter.

Guanidine Carbonate,

$$(H_2N \cdot CNH \cdot NH_2)_2H_2CO_3,$$

is an organic carbonate of approximately the same strength as sodium carbonate. It is soluble in water but only slightly soluble in alcohol and acetone. Its advantages over other organic amines are that it is entirely stable under normal conditions; it is a strong base; it does not cause yellowing of cosmetic emulsions; and it has a combining weight of 90. In combination with stearic acid, mineral oil, and waxes it forms stable water-in-oil emulsions.

In making soaps, the fatty acids should be warmed with an equal volume of alcohol before adding guanidine carbonate, to prevent thickening and foaming. These soaps are soluble in alcohols, benzene, carbon tetrachloride, ethylene dichloride, naphtha, and water. The properties of the soaps indicate their use as detergents, dry-cleaning soaps, general emulsifiers, and possibly in cosmetics.

Diethylaminoethanol, Ethylethanolamines 161

These amines produced by Pennsalt, form soaps with oleic acid, which are suitable for self-polishing emulsion for floor wax, cutting oils, automobile and furniture polishes, emulsion paints, and cosmetic creams.

To illustrate the use of three of the above amines together with a comparison of results, we quote from a letter received from Commercial Solvents Corporation:

> It will be necessary for you to make adjustments in our typical formula depending on the types of waxes used and your object in obtaining special properties in the finished product.

WAX-RESIN FORMULA FOR EVALUATING EMULSIFYING AGENTS

Carnauba #3 N.C. Refined	30.0 g
Emulsifiable Resin[1]	15.0
Oxidized Microcrystalline Wax[2]	15.0
Oleic Acid	Varied
Amine	Varied
Water	q.s. to make 500 g

1 Durez 219 (Durez Plastics & Chemicals)
2 Crown 23 (Petrolite Corporation)

Numerous emulsions were prepared to determine the optimum

concentration of the three most popular emulsifiers. The optimum concentrations used in the test formula are shown below.

OPTIMUM CONCENTRATION OF EMULSIFIER IN TEST FORMULA

	AMP		TEA		MORPHOLINE	
	Grams	% Wt*	Grams	% Wt*	Grams	% Wt*
Amine	4.54	7.56	8.24	13.72	7.10	11.82
Oleic Acid	7.20	12.00	7.80	13.00	9.00	15.00

* Based on wax-resin content (12% by weight)

Table 7:1 gives some of the physical properties, using the optimum concentrations listed of emulsifiers with the typical wax formula.

TABLE 7:1

PHYSICAL PROPERTIES OF WATER–WAX EMULSIONS

	AMP	TEA	MORPHOLINE
Stability at 125°F	Stable	Stable	Stable
Freeze-thaw	Satisfactory	Satisfactory	Satisfactory
Leveling on TOTL	Excellent	Excellent	Excellent
Gloss	15	14	15
Transmission, %	68.8	49.7	61.5
H_2O-Resistance, 24 hr	Good	Poor	Excellent
Removability, 24 hr	100	100	20
Tack, 24 hr	Fair	Poor	Fair
Abrasion, 24 hr	Fair	Poor	Fair

ADDITIVES AND BUILDERS FOR SOAP AND DETERGENTS

There are a number of compounds used in soap manufacture and added to synthetic surfactants to facilitate cleansing action. Most of them are more or less alkaline, so that they neutralize acidic substances which may be associated with soiled surfaces or materials to be cleansed. Many act as sequesterants for lime, magnesia, and iron. Moreover, these additives are not confined to alkalies alone.

Silicates

Sodium Silicate is one of the important additives to soap. It serves as a water softener as well as a detergent. In soaps which contain free alkali as caustic soda, the silicate combines with it and modifies its corrosive effect. It acts as an antioxidant for the soap itself, and it is occasionally added to white floating soap, in amounts up to 1% (N Brand).

Sodium silicate inhibits corrosion in the cleaning of aluminum and tin plate.

Other silicates have found place as ingredients of laundry and dish-washing detergents in combination with the various surfactants. Potassium silicate is an ingredient of heavy-duty liquid household detergents. Many of the uses of the various silicates are illustrated in formulas cited in this book, under appropriate headings. (Table 7:2)

TABLE 7:2

ANALYSES OF SILICATES OF SODIUM AND POTASSIUM*

Product	*Ratio* Na_2O; SiO_2	% Na_2O	% SiO_2	H_2O	*P.Q. Name*
Sodium Metasilicate 5 H_2O	1:1	29.5	28.7	41.7	Metso Granular
Sodium Metasilicate Anhydrous	1:1	51.0	45.5	0.5	,, Anhydrous
Sodium sesquisilicate Hydrated	3:2	36.7	24.2	38.1	,, 99
Sodium orthosilicate	2:1	60.8	27.5	9.5	,, 200
Dry 65-Mesh	1:2	27.5	55.0	17.5	GD silicate
Liquids:					
N-Brand	1:3.22	8.9	28.7	Sp. Gr. 1.394 abt. 41 Bé	
	K_2O: SiO_2	K_2O			
Potassium Silicate Kasil #1	1:2.50	7.9	19.8	Sp. Gr. 1.244 abt. 29 Bé	

* Philadelphia Quartz Co.

Soda Ash, Na_2CO_3, is the most common alkaline additive to laundry soap. It constitutes the major portion of cheaper soap powders. It is a good cleanser. It saponifies and neutralizes fatty acids present in soiled laundry and it is a water softener.

Phosphates

Trisodium Phosphate (TSP), $Na_3PO_4 \cdot 12\ H_2O$, also known as *Trisodium Orthophosphate*, is an alkaline compound which finds considerable use as a general cleanser for painted surfaces. For that purpose it should be used in very dilute solution. It has emulsifying action on animal, vegetable, and mineral oils. It also serves as a water softener. It is used in soaps, gritty cleansers, and in synthetic detergents.

Chlorinated Trisodium Phosphate (Blockson) has 3.7% of available chlorine. Its analysis is:

Sodium hypochlorite	>3.25
$Na_3PO_4 \cdot 12\ H_2O$	91.75
Sodium chloride	<5.00

It can be used in gritty cleansers as the major cleaning agent and at the same time to introduce the bleaching effect of free chlorine. It is used in general sanitizing for utensils and equipment in the food and beverage industries; typical solutions are as follows:

50 ppm	available chlorine	0.75 oz in 3.5 gallons water
100 ppm	,, ,,	1.5 oz in 3.5 ,, ,,
200 ppm	,, ,,	3 oz in 3.5 ,, ,,

Tetrasodium Pyrophosphate (TSPP), $Na_4P_2O_7$, is a water softener and sequesterant; therefore, calcium and magnesium soap precipitates are avoided. When TSPP is used as a builder with soap or a synthetic detergent, the resultant detergency is greater than the effect of the TSPP and the soap or synthetic, acting independently.

Tetrapotassium pyrophosphate (TKPP), $K_4P_2O_7$, is similar in properties to the tetrasodium compound. It is used in heavy-duty liquid detergents, liquid soap formulations, and in cleaning compounds. It is sold as an anhydrous powder and also as a water-white viscous solution with a total solids content of 61.0 per cent. A 1%-solution has a *p*H of 10.4.

Sodium hexametaphosphate $(NaPO_3)_6$ (Sodium Polyphos, of Blockson). "An infinite number of true glassy phosphates can be formed within the molar range

$$1Na_2O:1P_2O_5 \text{ to } 1.34\ Na_2O:1\ P_2O_5$$

The chemical, physical, and structural properties of these glasses, which include sodium hexametaphosphate and sodium tetraphosphate $(Na_6P_4O_{13})$, are of such similarity that in practical utility, one may be substituted for another. Sodium Polyphos has a molar composition of $1.1Na_2O:1P_2O_5$, which is in the high P_2O_5 end of the glass-forming range. In sequestration, the calcium and magnesium ions are held so firmly in a soluble chemical combination that they will not react with any soap that is added, or formed by saponification of greases and oils during the cleaning operation." (Blockson Chemicals.)

According to R. D. Stayner,

> Among the phosphates, the order of decreasing effectiveness for sequestration of calcium ions is:
>
> Sodium hexametaphosphate (Calgon)
> Sodium tripolyphosphate
> Tetrasodium pyrophosphate
> Sodium phosphate

Sodium Tripolyphosphate, both anhydrous $(Na_5P_3O_{10})$ and hydrated $(Na_5P_3O_{10} \cdot 6H_2O)$ are important adjuncts to synthetic detergents and sudsing agents. Small amounts of STP will inhibit precipitation of calcium and magnesium salts in hard water when the temperature is raised or the alkalinity is increased. Another outstanding property of STP is its ability to deflocculate and suspend certain water-insoluble substances. In laundering operations, it disperses soils and helps to keep them in stable suspension, thus preventing redeposition of soil on the washed articles, even during the rinsing operation.

Table 7:3 shows the approximate composition of industrial alkalies, and provides at a glance comparative information on the most important alkalies used in the industry.

TABLE 7:3

APPROXIMATE COMPOSITION OF INDUSTRIAL ALKALIES
(Courtesy of Philadelphia Quartz Company)

Material	*Alkali Content as NA_2O*			*Water (H-O)*	*Other Major Constituent*	*pH, 1.0% Solution*
	Titratable		*Total*			
	Active[1]	Inactive[2]				
Metso Granular	28.3%	1.2%	29.5%	41.7%	28.7% SiO_2	12.4
Metso Anhydrous	49.0	2.0	51.0	0	47.0% SiO_2	12.6
Metso 99	35.3	1.4	36.7	38.1	24.2% SiO_2	12.6
Metso 200	59.0	1.8	60.8	9.5	29.0% SiO_2	12.9
Caustic Soda	75.6	0.4	76.0	22.0	None	13.1
Soda Ash	28.7	28.7	57.4	Nil	42.0% CO_2	11.4
Sodium Sesquicarbonate	13.7	26.9	40.6	19.9	38.5% CO_2	9.8
Sodium Bicarbonate	0.9[3]	36.0	36.9	10.6	51.5% CO_2	8.4
Trisodium Phosphate	10.0	8.8	24.1[4]	56.8	18.2% P_2O_5	12.1
Tetrasodium Pyrophosphate (Anhydrous)	8.0	15.0	46.1[4]	Nil	52.9% P_2O_5	10.2
Sodium Tripolyphosphate	4.3	16.9	42.0[4]	Nil	57.0% P_2O_5	9.4
Hexametaphosphate	0.0	1.97	30.4[4]	Nil	69.6% P_2O_5	6.5
Sodium Tetraphosphate	0.0	10.17	39.6[4]	Nil	60.4% P_2O_5	8.5

1 Titratable with phenolphthalein; varies slightly with concentration.
2 Titratable with Methyl Orange; varies slightly with concentration.
3 Theoretically no active alkali, but commercial product shows a little.
4 Note only part of total alkali is titratable; remainder is fixed and unavailable for neutralizing acids.

Antioxidants

The addition of antioxidants is important, to prevent deterioration and discoloration (see Chapter 6).[7]

For soap, the following are used: Sopanox (Monsanto), *o*-biphenyl biguanidine (Monsanto #326), diphenyl amine, *p*-tert-butylphenol, *p*-tert-amylphenol. These chemicals in concentration off 0.1% may be added to the oil before saponification. For *P.S. 600b Soap, Laundry,*

Low Titer, diphenyl amine, *o*-diphenyl biguanidine, *p*-tert-amylphenol and *p*-tert-butylphenol are acceptable as inhibitors of oxidation, and may be added to the extent of 0.05% according to the specifications. R. T. Vanderbilt Antioxidant #16, was found by this author to prevent oxidation and spotting of corn oil soap for a number of years.

Other additives suggested by various investigators are formaldehyde, sodium stannite, sodium stannate, thymol, carvacrol, sodium thiosulfate, benzil, hexamethylenetetramine, acetaldehyde, aniline condensate (U.S. Rubber), Agerite (Goodyear), aldol, allyl alcohol, aluminum, aluminum sulfate, anisol, benzylamine, benzyl benzoate, borneol, citronellol, coumarin, cymene, eucalyptol, glycine, hydrazine, *o*-toluidine, sodium salicylate, safrol, isosafrol, diamyl phenol and stearic hydrazide.

Hydroquinone monomethyl ether has been advertised as an antioxidant and stabilizer which is oil-soluble but only slightly soluble in water, high-boiling but low-melting, and retentive of good color.

Miscellaneous

Borax ($Na_2B_4O_7 \cdot 10\ H_2O$) is a mild alkali (1% solution *p*H 9) that has been used in connection with soap. Even when laundry bar soaps were widely sold, few soaps contained actual borax. Borax is a household utility and will continue to be so. One of the borax producers, however, has prepared a new composition which has considerable acceptance. It is sold as Borateem and is described as "Activated." Its composition is:

Sodium tetraborate 10 H_2O (borax)	98.4
Dodecylbenzene sulfonate	0.4
3,4,4′-Trichlorocarbanilide	0.2
Inert	1.0
	100.0

It is sold in cartons holding 3 lb 7 oz and 6 lb 14 oz.

Another use to which borax is put, is as a component of soap for dispensers. This is a good powdered hand soap for mechanics' use. The borax, which is the major component, furnishes abrasive effect and finally dissolves; (see Specification P-S-628*a*, under appropriate heading).

Puffed Borax is a new development of the American Potash & Chemical Corporation. By proper technique of heating the decahydrate ($Na_2B_4O_7 \cdot 10\ H_2O$), a product is made with varying amounts of water. That is, V-Bor ($Na_2B_4O_7 \cdot 5\ H_2O$), and Pyrobor ($Na_2B_4O_7$) without changing the ratio of Na_2O to B_2O_3. If the base chemical is heated rapidly, the escaping water explodes the crystal, like popcorn, thus puffing the particle to several times its original size. Puffed borax weighs 10 lb/cu ft; V-Bor has a density of 65 lb/cu ft.

It is possible to load the puffed product with most organic materials. Liquids can be applied directly on a moving bed of the puffed borax. Solids can be either melted or dissolved in a volatile solvent and applied as liquid, with subsequent evaporation of the solvent. A puffed product with a density of 4 lb/cu ft yielded a free-flowing material containing 90% by weight of trichloroethylene. All halogenated hydrocarbons are absorbed and yet the resulting products remain free-flowing. The puffed borax has good stability. It is not fragile and does not break down or powder in packaging and shipping.

The inclusion of puffed borax in detergents can increase bulk when that is desirable. Also because of its absorbency, it permits the use of liquid detergents without producing a soggy powder. It is instantly soluble, too.

Bentonite, or colloidal clay, has sometimes been used in bar soaps. It is assumed that it facilitates washing because of its colloidal form. It absorbs water and imparts an appearance of dryness in the presence of considerable moisture. It adds stiffness to the bar of soap, which would be less firm without it. It was a component of a liquid window cleaner (Glass-Wax), which is still on the market, but has lost favor, because of the residual powder left on the window frames and because of the introduction of aerosol window cleaners and other sprays.

Naphtha is used in bar soap as a supplementary cleaner. It has solvent action on grease and soil. Bar soaps may contain as much as 5% of naphtha. Today there are few naphtha soaps on the American market. We would, however, recommend it for use in those countries where washing machines are not yet in vogue.

Mineral Soap-Stock is another petroleum product which was used in rosin soap. This may be a light spindle oil or a soft petrolatum. The action of both types is similar to that of naphtha.

Low grade rosin soaps also used to contain marble dust or silica, which were merely make-weights and acted as abrasives.

Powdered detergents, such as are used for mopping tile and terrazo floors, contain *silica* or *seismotite* (volcanic ash), which facilitates the cleansing by its abrasive action. Volcanic ash does not precipitate so readily as powdered silica nor does it scratch so much, because of its friability. Because of its light weight it adds bulk to the product.

Powdered pumice is used in mechanic's hand soap and in powders. Obviously the pumice helps to remove the soil by its abrasive power. In paste soap, pumice can be distributed more easily than silica because it is much lighter and remains in suspension without settling to the bottom of the container while the hot mixture is congealing.

High grade powdered cleansers contain up to 85% of *feldspar*, which is softer than silica; 325-mesh may be used for cleaning windows, without scratching. This type of cleanser is used also for cleaning bath tubs, sinks, marble and tile surfaces. Some manufacturers prefer coarser abrasives because they work faster.

Low-grade powdered cleansers contain 95% of Silex, 1–2% of soda ash, and 1–2% of soap or even synthetic surfactant.

In milled toilet soap, up to 1% by weight of zinc oxide or titanium dioxide is added as whitener and opacifier.

Other additives, such as lanolin, lecithin, bacteriostats, etc., are discussed in other chapters. [8]

COLORING SOAP AND DETERGENTS

The tinting of soaps and detergents should be attended with care and precision. Not all colors are resistant to heat, light, and alkalinity. Consultation with a dye manufacturer is one of the best means of obtaining the proper color for any specific purpose. A color once adopted, should be used repeatedly and preferably from the same source, so that it is uniform in strength and tinctorial power.

The soapmaker, for the most part, uses water-soluble dyes. For toilet and industrial soaps, the so-called *technical purpose colors* are used, but for liquid soaps and shampoos, the certified D&C or FD&C colors are mandatory.

In using any type of dye, the soapmaker or compounder should make it a rule to dissolve one ounce of dye in hot water and dilute to one gallon. The dye solution should then be strained through muslin, to make sure that there are no undissolved particles which may form dark specks in the soap. Color solutions should be stored in glass or other non-corrodible containers. If solutions are to be stored for a considerable time, 0.1% of a suitable preservative such as sodium benzoate should be added. Some dye solutions become moldy upon long standing.

Should it be necessary to color boiled soap, the color solution may be added during the settling change, before the boiling is stopped. Low-titer soaps may be colored in the crutcher.

Test for Correct Shade

Milled toilet soaps are colored in the amalgamator. It is difficult to advise as to the exact amount of color needed. Obviously, this must be determined by experiment. To one pound of chips, an ounce or two of color solution is added, distributing the color over the chips in a shallow dish or on a paper and mixed with a spoon or spatula. The chips are then run through a meat chopper and finally compressed in the shape of a finished bar of soap, for evaluating the shade of color. If the color is too light, the experiment is repeated. After breaking up the soap into small particles, more color is added, again ground and compressed, until the desired color is achieved.

The amount of color used, should be noted and used as a base for calculating the quantity required for an amalgamator batch, which is about 200 pounds. It is advisable to use a little less color in the batch than is actually calculated from the experiment, especially if the experimental batch is not compressed by machine. The plodding and subsequent pressing, naturally produces a more compact bar, which would be considerably darker than that compressed by hand.

For a large batch of soap the requisite amount of color solution is poured over the chips from one end of the amalgamator to the other, as uniformly as possible, while the machine is running. In about 15 minutes, the color and perfume, if any, should be evenly distributed throughout the soap. Where small amounts of solution are required, as in pastel shades, it is good practice to dilute the dye still further to ensure more even distribution.

Mottled Coloring

Blue-mottled or red-mottled soaps are colored, respectively, with either ultramarine or vermilion. It has already been mentioned (see *Mottled and Variegated Soaps*) that for real Eschweger soap, the color, dispersed in oil or mixed with one-half gallon of water containing 0.1% of caustic soda, based on weight of soap, is added just before the boiling is stopped. In imitation mottled soap, however, as produced by the cold process, the pigment may be dispersed in a portion of the oil to be used in the batch and it is added to the frame after drawing from the crutcher. The colored oil is poured over the top of the frame of soap and then crutched in by hand with a wooden crutch made by attaching a 2 × 4-inch block of wood, 8 inches long, to a handle long enough to reach to the bottom of the frame. The color is distributed by several rapid up-and-down strokes. This produces a variegated soap.

To color half-boiled soap, made with either potash or soda, the color solution is added after saponification is complete, when the dye may be mixed in thoroughly.

Liquid Soap

Liquid soap should be colored immediately before running to the refrigerator tank, before filtration. One of the most popular colors for this purpose is D&C Yellow No. 8, which imparts a brilliant fluorescent yellow to the liquid. Perfume, if any, should be added at the same time.

Powder and Flakes

At times it is necessary to color soap powder, especially when very light-colored fats have been used in the soap base. Yellow color may be added to the soap slurry before addition of the soda ash. This imparts an appearance of richness to the finished product.

In every case, care should be exercised that the finished soap does not produce colored suds. Toilet soap should not be colored so deeply that it will tint a wash cloth or stain bath room fixtures.

For specialty soaps, such as bluing soap flakes for laundry use, the dye should be added in small batches before running through the dryer. In most cases, however, the blue is added in the amalgamator, and then run through the mill, which imparts a gloss to the finished flakes. One manufacturer uses an embossed roller to produce a blue flake in a diamond shape.

Fluorescent Dyes

The development of the so-called "sunlight ingredient" in soaps and detergents has in a measure supplanted bluing. Colorless, blue-fluorescent dyestuffs such as stilbene derivatives, triazines, or modified umbelliferones, absorb ultraviolet radiations and reemit the energy in the form of visible blue radiations. These dyes are added both to soaps and to detergents. Sometimes they are used together in powdered detergents.

For detergents, both powder and liquid, the following Pylam Products' colors are quite efficacious:

Pylaklor S-500, Blue; S-505, Pink; S-507, Yellow.

Most brighteners—sometimes called white dyes— are variations of two basic types—diaminostilbenedisulfonic acid derivatives for cottons, and naphthyl triazole stilbene for cottons and nylon and also for bleach stability. They are used at levels of about 0.2%, based on the weight of the product. The dyes are used in the proportion of two parts cotton brightener to one part of cotton and nylon brightener.

Some brighteners are produced for liquid detergent and also for cationic laundry softeners. We suggest consultation with the dyestuff manufacturers, to obtain the proper dye for a specific use.[9]

PERFUMING SOAPS AND DETERGENTS

In large-scale manufacture, the perfumes to be used in soaps are blended at the soap plant. Natural oils as well as synthetics may be used. In producing a really fine blend, the perfumer's art plays a very important role.

Perfume is required to cover the natural odor of the soap or detergent. It should be pleasant and lasting. In some cases it must be heat-resistant to withstand the temperature of hot soap, while it is being mixed in the crutcher. The finished product must have a long shelf life and should not oxidize while awaiting sale.

This implies a thorough knowledge of essential oils, synthetics isolates, fixatives, etc. This is work only for a specialist in perfumery. Only the large manufacturers can afford such an expert, who is employed for this sole purpose.

For small-scale industry, perfuming may be more efficiently and economically carried out by consultation with the established perfume houses who can provide the correct type of perfume for each purpose. In that way, large inventories of many ingredients which make up a blend, will be avoided and better all-around results will be assured.

For perfuming laundry soaps, the cheaper essential oils or synthetics may be used, at the rate of 1–3 pounds for a 1200-pound frame. Perfume is added to this soap merely to offset the odor of the soap itself. Citronella, pine, sassafras, and some terpenes are standard odors for this purpose. This may apply to some industrial soaps also.

Some specialty soaps and flakes used in home laundering are delicately scented so as to impart the odor to the clothes being washed. These sell well for this sole reason.

In perfuming milled toilet soap, 0.5–1% of a pleasant bouquet is adequate. The perfume is added to the flaked soap in the amalgamator so that very little of the fragrance is lost through volatilization. This permits the use of very delicate odors.

For perfuming soap shampoos, the desired effect may be had by adding perfume at the rate of one-half ounce to the gallon. Where concentrates are produced for later dilution, more of the perfume will be required, so that the finished product will have the requisite amount of perfume.

Toilet goods and soap for men should not be so highly scented that the fragrance clings for a long period.

As there are so many types of detergents, it is difficult to generalize on the type of perfuming. Needless to say, the alkylaryl sulfonates require more perfume than many other types. Occasionally the heavy kerosine and sulfurous odor remains on the fabric after laundering. This must be overcome at all costs.

Detergents based on coconut oil, such as the lauryl sulfates, retain some of the fruity odor of the coconut and consequently do not require heavy perfuming. Although much of the detergent production goes into home laundering and dish-washing these products require careful attention to the perfuming. It is best to consult the experts on such problems.

It is a habit of people to smell soap or detergents before using. If the initial experience is pleasant, repeated sales may be expected. On the other hand, if the odor is offensive, no matter how fine the quality of the product may be, it will not enjoy repeated sale.

One has only to recall that cresylic acid as an odorant for toilet soap, has been succeeded by finely perfumed soap, which contain odorless bacteriostats.

MEDICAL USES OF SOAP

"Regardless of the kind of soap or other detergent used, and regardless of the frequency and thoroughness with which the scrubbings are performed, the bacterial flora can never be reduced to zero." So says Dr. Morris Fishbein in *Medical Uses of Soap*.[10] Other authorities have stated that soap has definite bacteriostatic and bactericidal effect. Evidence proves that in the manufacture of disinfectants, certain soaps used in the emulsification of either pine oil or phenolic ingredients affect the phenol coefficient. An example is the use of rosin soap in the preparation of pine oil disinfectant, which normally yields a product with a phenol coefficient of 3. When coconut or castor oil soaps are used, however, the coefficient will be increased to 4 or 5. Similar results have been obtained with phenolic disinfectants. A case in point is the admonition by Monsanto, to use saturated oil soaps in connection with Santophen I.

The U.S. Pharmacopoeia recognizes two types of soap: *sapo mollis* and *sapo durus*. The former is a soft soap, consisting of unsaturated oils saponified with potash. The latter consists either of olive oil or its equivalent, but saponified with caustic soda.

Sapo mollis is used in making *Tincture of Green Soap*, long used by physicians for scrubbing up before surgery. It may also be used in cases of acne and other skin disorders. For physicians' scrub-up, soap has been to a large extent displaced by detergent lotions, containing hexachlorophene or other bacteriostats. *Sapo durus* is the basis of soap liniments, but it is little used nowadays.

There are a great many proprietary medicated soaps sold. These contain such ingredients as mercuric iodide, phenylmercury compounds, hexachlorophene, bithionol, sulfur, cresylic acid, pine tar, lanolin, resorcinol, hexylresorcinol, chlorocarvacrol, and the more modern bacteriostats, and germicides such as tetramethylthiuram disulfide, trichlorocarbanilide, tribromosalicylanilide, tetrachlorosalicylanilide, trichlorosalicylanilide, dichlorobenzoxazolinone (a fungistat). Sometimes two of these are used together because it is thought that they are synergistic. Elsewhere in the book, these are mentioned together with the proprietary products in which they are used.

Veterinary soaps may contain extracts of derris and pyrethrum for killing fleas on dogs.- The treatment is said to last three months.

As some of the bactericides mentioned are said to be substantive to the skin and to fabrics, continued use actually reduces skin flora.

FEDERAL SPECIFICATIONS FOR SOAP AND SOAP PRODUCTS

P-S-311a: Type I
Grit Scouring Powder for Marble Floors

COMPOSITION

	Class 1		*Class 2*	
	min	max	min	max
Matter volatile at 105°C	–	8.0%	–	10.0%
Anhydrous soap and/or active salt-free synthetic detergent	1.0	7.0	2.0	10.0

Free alkali as NaOH	–	0.1	–	0.1
Insoluble siliceous matter	60.0	95.0	60.0	95.0
Bleaching strength:				
Available oxygen, as sodium perborate·$4H_2O$;				
or				
Available chlorine calculated as sodium hypochlorite	–	0.6	–	0.6
Alkaline salts, carbonates, phosphates	–	8.0	2.0	20.0
Fineness of Insoluble siliceous matter:				
% retained on No. 60 US Standard Sieve	1.0%		1.0	
No. 100 US Standard Sieve	1.0		1.0	
No. 200 US Standard Sieve	5.0		20.0	

Suggested Formulas

	Class 1	*Class 2*
Swift's or equal 90%-active synthetic detergent	6.0%	10.0%
TSP–12 H_2O	5.0	15.0
Soda Ash	5.0	5.0
Silex (200-mesh)	84.0	70.0
	100.0%	100.0%

In Class 1, if the chlorine type of bleach is desired, 0.5–1% of Monsanto's ACL 59 (or equivalent) may be incorporated. If perborate is preferred an equal amount of sodium perborate may be added. Perfume is optional.

P-S-320a
Scouring Powder, Glass Cleaning

COMPOSITION

	min %	max %
Matter volatile at 105°	–	4.0
Total alkali and matter insoluble in alcohol calculated as Na_2CO_3	–	5.0
Total alkali, as NaOH	–	0.1
Anhydrous soap and/or active synthetic salt-free detergent	4.0	–
Insoluble siliceous matter	85.0	93.0

Suggested Formula

200-mesh feldspar	90.5%
Soda ash	4.5
Swift's 90% detergent flake or powder (or equal)	5.0
Total	100.0%

Procedure:

Blend the ingredients in a ribbon-type mixer. For retail trade, 0.1 to 0.2% of perfume, such as pine oil, isobornyl acetate, or diphenyl oxide may be added.

P-S-560

Soap, Automobile and Floor, Liquid and Paste

TYPE I, LIQUID: For general cleaning and in undiluted form, for detecting gas leaks (as in oxygen systems).

TYPE II, PASTE: *Class 1*, for general cleaning; *Class 2*, for use as tire-tube lubricant.

COMPOSITION

	TYPE I		TYPE II			
			Class 1		*Class 2*	
	% wt		% wt			
	min	max	min	max	min	max
Moisture (toluene distn.)	–	–	–	55.0	–	60.0
Total matter insoluble in alcohol	–	0.5	–	1.0	–	–
Free alkali as KOH	–	0.05	–	0.1	–	0.1
Free acid, as oleic	–	0.1	–	0.2	–	0.2
Alkaline salts, as K_2CO_3	–	0.1	–	0.2	–	0.2
Matter insoluble in distilled H_2O	–	0.1	–	0.2	–	0.2
Chloride, as KCl	–	0.3	–	0.5	–	0.5
Unsaponified, unsaponifiable	–	0.8	–	1.8	–	1.8
Anhydrous soap, as K soap	20	–	43.0	–	43.0	–
Total Na compounds, as Na_2O	–	0.2	–	0.5	–	0.5
Glycerol	–	1.8	–	4.0	3.0	–
Rosin acids	–	1.0	–	2.0	–	2.0
Sugar	–	none	–	none	–	none
Numerical Values:						
Iodine No. (Wijs) of mixed F.A. from soap	80	150	80	150	–	–
Acid No. of mixed F.A.	190	205	190	205	190	205

Computation:

TYPE II: The percentage of moisture shall be computed and reported on the soap as received. The percentages of all other constituents shall be calculated and reported on an assumed moisture of 50 percent. To make these calculations, multiply the percentages of constituents other than moisture by a factor (F) derived as follows:

$$F = \frac{100 - 5}{100 - \%\,\text{moisture}}$$

The soap required under this specification, is a potash vegetable oil soap, in which the soapmaker has the choice of using crude corn oil, crude soya bean oil, semi-refined cottonseed oil with optional addition of 5–10% refined tall oil, which may contain about 10% of rosin acids. The choice is usually governed by the market price of the oils. Where color is not a factor, the maker sometimes uses acidulated vegetable soapstock, if it is obtainable. The latter is usually in the hands of the large operators who convert the lower grades by distilling the fatty acids and thereby derive greater revenue.

The simplified formula would be:

Crude corn oil	40 lb
45% caustic potash	17
Water	43
Total	100 lb

Procedure:

The corn oil is pumped to a steam jacketed kettle, water is added a little in excess of that shown above, for there should be allowance for evaporation during the boiling. The two are heated to boiling, when the potash is added little by little as it is absorbed by the oil. In no case should the potash be added so that there is a considerable excess. Under such condition the oil will not take up the potash and boiling may proceed for hours.

In a small kettle—say 100-gallon capacity—a perforated steam coil may be placed directly in the kettle to supply steam and agitation and at the same time to replenish the evaporated water. When all the potash has been absorbed, and the soap boils evenly and appears to be translucent, a test is made for completeness of saponification. This may be done by simply dissolving a little soap in hot distilled water. The solution must be clear. If not, the boiling may be continued. If a sample of the soap, after continued boiling still clouds in a distilled water solution, the soap may require a small amount of additional potash.

Before adding more potash, however, a test for free caustic should be made. If the soap is weak, potash may be added cautiously and boiled again. If there appears to be sufficient alkali, or an excess, the soap may not have been boiled sufficiently. With a little experience, the maker will learn when to add potash, a little oil or fatty acid, and water. When the batch is finished, it should be tested for free caustic and moisture, and proper adjustment should be made. A little soap should be taken out to cool off. Then one can make appraisal as to whether the soap has the proper consistency and appearance.

Should the soap be too alkaline, it might be well to have on hand some vegetable fatty acid for quick neutralization, although this can be accomplished with the same oil as used in the batch. In either case, the last addition must be boiled in, so as not to show specks of soap that are darker than the batch itself.

At the option of the maker, a little inexpensive perfume may be added, although for federal specification, perfume is not required. For janitorial use, some prefer a pine or sassafras odor.

P-S-563b

Soap, Built, High-titer, Powdered

This is a soap suitable for high-temperature laundering of all types of cottons and linens, but not silk and woollens.

COMPOSITION

	Min %	Max %
Moisture and matter volatile at 105°C	–	16.0
Sum of free alkali and total matter insoluble in alcohol (Int. Amendment)	25.0	50.0
Matter insoluble in water	–	1.0
Titer of mixed F.A. derived from soap	39°C	–
Material retained on No. 12 sieve	–	1.5
Rosin based on finished product	–	none
Anhydrous soap	50.0	–
Free alkali, as NaOH	–	16.0

Computation:

The percentage of volatile matter at 105°C shall be calculated on the basis of soap as received, but all other constituents shall be calculated on the basis of material containing 16% of volatile matter at 105°C.

This soap is prepared by full-boiled process, with tallow and caustic soda. The crutching formula is as follows:

Kettle soap	800 lb
N-Brand silicate	250
37°Bé caustic soda	10
Soda ash	140
Perfume, such as Javal (Felton) optional	

It is possible that a little hot water will be necessary to plasticize the mixture in the crutcher. The soap is then run on the dryer, so regulated as to yield a chip of about 16–17% moisture. Alternatively, the mixture may be run into frames, and cooled to solidify it. Then it may be slabbed, barred, racked off to dry and prepared to be chipped on the chipping wheel. This method is for plants that do not have a chip dryer. The chips may be dried on trays to proper moisture and then ground in the mill.

P-S-571c
Soap, Scouring, Cake Form

Soap, scouring, cake form, shall be made from high-quality soap and abrasives, uniformly mixed. Each cake shall weigh not less than 9 oz nor more than 10 oz.

COMPOSITION

	Type I % weight		Type II % weight	
	min	max	min	max
Moisture and matter volatile at 105°C	–	5.0	–	6.0
Anhydrous soap	6.0	10.0	8.0	12.0
Alkaline salts, as Na_2CO_3	–	1.5	–	4.0
Free acid, as oleic	–	0.5	–	0.5
Free alkali, as NaOH	–	0.1	–	0.1
Sugar or other foreign matter	–	none	–	none
Material insoluble in water	–	remainder	–	remainder

Abrasiveness:

Type I shall not scratch glass; therefore it should consist of a softer abrasive, namely 200-mesh feldspar.

Type II, which is for coarser work, shall consist of ground quartz. The silex must pass 99% through 100-mesh sieve.

These products are made in a horizontal dough-mixer, into which open steam may be admitted. To produce a 1000-lb batch, 850 pounds of feldspar is dumped into the mixer and 150 pounds of a 50% slurry of soap is added. The whole is mixed while live steam is admitted through jets at the bottom of the mixer. The finished mass should be uniform hot dough, showing no specks of soap. The soap is now run into a frame and pounded down well, so as to eliminate air pockets as far as possible.

This mass is allowed to cool several days, when it will have caked quite hard. The soap is slabbed and cut into brickettes which are placed on racks to dry, preferably in the air at room temperature. When the drying is forced, the bars are apt to develop cracks. Instead of being cut in the ordinary way, there is a special cutting press for grit soap, which cuts each slab into small cakes and imprints them at the same time.

Type II is made in a similar manner, using 100–140-mesh silex. To 800 pounds of silex, 170 pounds of 50% soap slurry and 30 pounds of soda ash are added and mixed as described.

P-S-576c
Soap, Grit (Hand Cake)

Hand Grit Soap shall consist of soap, soap additives, and finely ground insoluble siliceous material, in a compressed cake. The product may be mildly perfumed.

COMPOSITION

	min %	max %
Moisture and matter volatile at 105°C	–	25.0
Total alkalinity, matter insoluble in alcohol; alkaline salts, as Na_2CO_3	–	1.0
Free alkali, as NaOH	–	0.1
Free fatty acids, as oleic	–	0.5
Insoluble siliceous material	22.0	35.0
retained on #100 sieve	–	2.0
retained on 200-mesh sieve	–	10.0
Rosin	–	5.0
Sugar	–	none
Anhydrous soap	35.0	–

This type of soap, usually known as *pumice hand soap*, is generally based on regular toilet soap and it can be produced by milling in the pumice.

Suggested Formula

Toilet soap chips or scrap	65.0 lb
Pumice	35.0
Perfume oil, sassafras or Javal or equal	1.0
	101.0 lb

The moisture content will be governed by the amount of moisture in the soap chips or scrap.

As the specification does not specifically call for pumice, 200-mesh silex may be used. Also, where special soap base is produced for this purpose, about 15% of refined tall oil containing about 25% rosin may be used.

Alternate Method

There is an alternate method for producing a soap of this type, using a semi-boiled soap made with crude coconut oil. The formula is:

FFF pumice	30 lb
Crude coconut oil	40
Caustic soda, 37°Bé	23
Water	6
Perfume	1
	100 lb

Heat coconut oil to 170°F in the crutcher and while agitating, sift in the pumice. Then gradually add the lye and the water in a thin stream, allowing the heat to rise to 180°F. If the soap at the end of the saponification appears to be too thick, add about 5% of water and crutch in well. Test for free caustic; if found satisfactory, the soap may be run into a frame to cool. (At the end of the saponification the soap should be smooth and slide off the trowel like settled soap.)

When it is cold, slab and cut the soap before it becomes too hard. After surface drying, the bars may be pressed.

In this formula too, silex may be used in place of pumice, but it is more difficult to keep the heavy mineral from sinking to the bottom of the frame. With a little experimentation, this can be accomplished.

P-S-577
Soap, Grit, Paste and Powder

Type I: Hand Soap, Grit paste
Type II: Hand Soap with Mineral Abrasive
Type III: Hand Soap with Vegetable Abrasive and Lanolin

COMPOSITION

Percent by weight

	Type I		Type II		Type III	
	min	max	min	max	min	max
Matter volatile at 105°C	–	55.0	–	5.0	–	12.0
Alkaline salts, as Na_2CO_3	–	3.0	2.0	5.0	–	7.7
Free alkali, as NaOH	–	0.1	–	0.1	–	0.1
Free acid, as oleic	–	0.5	–	0.5	–	1.0
Anhydrous soda soap	8.0	–	17.0	–	24.5	–
Lanolin USP anhydrous	–	–	–	–	3.0	–
Matter insoluble in water	25.0	50.0	60.0	76.0	45.0	55.0
Rosin	–	–	–	5.0	–	none
Sugar	none		none		none	
Fineness of insoluble siliceous matter:						
% retained on 40 sieve	–	1.0	–	1.0	–	–
60 sieve	10.0	20.0	–	5.0	–	–
80 sieve	30.0	45.0	–	–	–	–
100 sieve	35.0	55.0	–	30.0	80.0	–
200 sieve	60.0	–	–	60.0	–	–
Fineness of finished soap,						
% retained on 20 sieve*	–	–	–	–	–	10.0
100 sieve	–	–	–	–	80.0	–

* Particles shall be soft and friable, not solid or hard. Bentonite shall not be included in the product.

Suggested Formula to make 100 lb

Type I, Paste Hand Soap

Laundry soap flakes	10.0 lb
Soda ash	2.5
Pumice (100-Mesh)	32.5
Perfume (Artificial Sassafras)	0.1
Water	54.9
	100.0 lb

Dissolve the soap base in hot water at 150°F and strain through a screen to remove undissolved particles. Add soda ash and stir into a smooth gel without froth. Around 135°F, add pumice and perfume, and stir until uniform. Run soap into cans or jars, at a temperature that will allow the soap to gel before the pumice sinks to the bottom. The temperature can be best determined experimentally. At summer temperatures it may be necessary to use fans to cool the soap in the containers, to speed congealing.

NOTE: If the soap seems to to be too hard or brittle, reduce the amount of soda ash. That will thin the gel.

Type II, Powdered Hand Soap

(Mineral abrasive)

Powdered laundry soap	20.0 lb
Soda ash	4.5
Pumice (100-mesh)	75.5
Perfume	0.1
	100.0 lb

Blend all in a ribbon-type mixer. When the entire mix is uniform, it can be packed.

Type III, Powdered Hand Soap

(Vegetable abrasive and lanolin)

A.	Powdered laundry soap	32 lb
	Soda ash	7
	Corn cob meal	52
B.	Lanolin	3
	Kerosine (deodorized) or	6
	Chlorothene NU (Dow)	6

Prepare a solution of B. Place ingredients of A into a ribbon-type mixer and while the batch is mixing, pour the solution slowly over the entire length of the mixer. When the batch is uniform it may be spread out on a concrete floor near a window, to evaporate the solvent. The mass can be turned over with a shovel several times. Perfume may be added if desired.

NOTES: The specification states that Type III shall be a uniform, free flowing, non-stratifying mixture of clean corn meal, corn cob meal, or a combination thereof, thoroughly saponified soap, and lanolin, in powdered form. The mixture shall contain no rosin, sugar, pine oil, nor mineral abrasives such as lava, pumice, sand, or quartz. Corn meal used shall be degerminated. (Amendment 2: dated May 12, 1961).

If this mixture does not yield sufficient suds in accordance with test method as in specification, powdered toilet soap should be used in preference to laundry soap.

If unprocessed corn meal is used, the free fatty acid content (calculated as oleic acid) shall be 2.5% maximum.

If the corn meal used is not degerminated, then the corn meal shall be protected against oxidation by the addition of an efficient anti-oxidant thoroughly incorporated into the corn meal. The minimum anti-oxidant content shall be 0.01% of the corn meal. The antioxidant shall be added with the liquefied lanolin or suitable carrier so that the anti-oxidant is thoroughly and uniformly dispersed into the corn meal. The type and amount of anti-oxidant used, shall be satisfactory to the procuring activity.

P-S-578
Soap, Laundry, Built, Low-titer, Powdered

The specification covers a low-titer soap, built with alkaline salts, suitable for use at low temperatures.

COMPOSITION

	% min	% max
Moisture and matter volatile at 105°C	–	16.00
Sum of free alkali and total matter insoluble in alcohol (alkaline salts, as Na_2CO_3)	21.00	40.00
Free alkali calculated as NaOH	–	0.20
Matter insoluble in water	–	1.00
Rosin	–	none
Anhydrous soap	50.00	–
Titer of mixed fatty acids	–	26.0°C
Material retained on #12 sieve	–	1.50
Material passing through #140 sieve	–	20.00

Flakes from dryer must be ground to meet sieve tests.

The fat to be used in this soap may consist either wholly of crude coconut oil or of a whole-boiled soap, consisting of 20% of white grease and 80% of coconut oil; or even vegetable F.A. of the proper titer and color.

A 1200-lb crutcher batch would contain:

Boiled soap	725 lb
Soda ash	150
Sodium silicate N-Brand	325
	1200 lb

When this is dehydrated to 16% of moisture, the product should meet the specifications given.

P-S-580a
Soap, Laundry, Chip

COMPOSITION

	% min	% max
Moisture and volatile at 105°C	–	10.00
Sum of free alkali or free F.A., total matter insoluble in alcohol, and NaCl	–	6.00

Free alkali, as NaOH	–	0.20
Free acid, as oleic	–	0.20
Matter insoluble in water	–	1.00
Anhydrous soap	85.00	–
Titer of mixed F.A. prepared from soap	39°C	–
Rosin	–	–

The soap as received shall not have an objectionable odor. It shall not leave an objectionable odor on objects after washing with it and rinsing thoroughly in hot water.

This is a whole boiled soap made with tallow, suitable for washing, cleaning, and scouring purposes with soft water, where the presence of alkaline salts is not desirable.

A crutcher batch consists of the following:

Kettle soap	1150 lb
Soda ash	25
N-Silicate	25
	1200 lb

When this is dried to the correct moisture, it will meet the specifications.

P-S-581b
Soap, Laundry Chip (Rosin Type)

The specification covers rosin-type chip laundry soap for use in heavy-duty laundering, such as heavily soiled occupational clothing, where high temperatures are required for washing.

COMPOSITION

	% min	% max
Moisture and volatile matter at 105°C	–	15.00
Sum of free alkali or free F.A., total matter insoluble in alcohol and NaCl	–	12.00
Free alkali, as NaOH	–	0.50
Free acid, as oleic	–	0.50
Chloride, as NaCl	–	1.00
Rosin	–	20.00
Anhydrous soap	72.00	–
Titer of mixed F.A. prepared from the soap	29°C	

This should be a whole-boiled soap consisting of about 80% of yellow house grease and 20% of rosin, which will satisfy the specifications as to rosin and titer.

A 1200 lb crutcher batch consists of the following:

Kettle soap	1000 lb
Soda ash	40
N-Brand silicate	160
	1200 lb

When this soap is run through the dryer at about 15% moisture content, it will meet the specifications given.

P-S-583b

Soap, Laundry (granulated, rosin-type)

COMPOSITION

	% min	% max
Moisture and matter volatile at 105°C	–	10.0
Sum of free alkali or free F.A., total insoluble in alcohol, and NaCl	–	14.0
Chloride as NaCl	–	1.0
Rosin	–	20.0
Anhydrous soap	75.00	–
Residue retained on #12 sieve	–	2.00
Titer of mixed F.A. prepared from the soap	29°C	–

This product can be made by grinding the chips of preceding (P-S-581b). In grinding, the soap will lose around 5% of moisture and will then meet the specifications.

P-S-591e

Soap, Laundry, Bar

COMPOSITION

	% min	% max
Moisture and volatile at 105°C	–	36.00
Sum of free alkali or free acid, total matter insoluble in alcohol, and NaCl	2.00	11.00
Free NaOH	–	0.20
Free acid, as oleic	–	0.50
Rosin	–	25.00
Anhydrous soap	52.00	–
Titer of mixed fatty acids prepared from the soap	29°C	–

A crutcher batch may be compounded as follows:

Kettle soap	950 lb
Soda ash	50
N-Brand silicate	200
	1200 lb

It is customary to add a pound of citronella oil to a frame. This formula is calculated to meet the specifications.

P-S-596c

Soap, Laundry, Powdered

This specification covers a white or light amber powdered laundry soap, suitable for use in high temperature laundering of soiled cotton fabrics and for general cleaning with soft water. Required here is the same type of soap as in specification P-S-580a, except that it shall be in such powdered form, that a maximum of 1.5% will be retained on the #12 sieve and contain a maximum of 6% of moisture. This can therefore be obtained by grinding the chips.

P-S-600b

Soap, Laundry, Low-titer

(For low-temperature washing)

The specification covers a neutral soap primarily for use in cleaning wool and other fabrics, finished surfaces, leather, and other articles where comparatively low-temperature washing solutions are indicated.

COMPOSITION

	% min	% max
Matter volatile at 105°C	–	7.00
Sum of free alkali or free acid, total insoluble in alcohol, and NaCl	–	8.00
Water-soluble sodium silicate	2.00	4.00
Free alkali, as NaOH	–	0.20
Free acid, as oleic	–	none
Matter insoluble in water	–	1.50
Rosin	–	none
Sugar	–	none
Copper, ppm	–	10.0
Unsaponified saponifiable matter	–	1.00
Anhydrous soap	86.00	–
Titer of mixed fatty acids prepared from the soap	–	18°C
Iodine number of mixed fatty acids prepared from the soap	80	–
Organic oxidation inhibitor [a]	0.05	–

[a] Acceptable inhibitors of the following types shall be used unless otherwise specified: Diphenylamine,* orthodiphenylbiguanidine; *p-tert* amyl phenol;** *p-tert* butyl-phenol.*
*Suppliers: 5; 69

This is a boiled soap made from distilled oleic acid of good color. A 1200-pound crutcher batch is made up as follows:

Kettle soap	1100 lb
Soda ash	40
Sodium silicate	60
Citronella	1
Antioxidant, as indicated	

When the soap is dehydrated to a moisture content of 7%, the chips or powder will comply with the specifications.

P-S-609a

Soap, Saddle

The specification covers one type of saddle soap, that is used to clean and to a certain extent, preserve and extend the useful life of such items as shoes, handbags, luggage, leather coats, upholstery, gloves, belts, sports equipment, etc.

The soap shall be a non-flowing smooth paste, over a temperature range of 5°C to 38°C with no separation of vehicle and in proper form for application, so that it may be readily taken up with a brush or cloth without crumbling.

This may be either a half-boiled or a whole-boiled soap.

The fats used may consist of 25% Niger palm oil and 75% house grease.

A crutcher batch, made from whole-boiled soap, will consist of:

Kettle soap	400 lb
Sodium silicate	5
Hot water	717
Beeswax	36
Neatsfoot oil	36
D-C Silicone 555	6
	1200 lb

Procedure:

Run the soap into the crutcher, add hot water and silicate. Stir until a nice smooth slurry is formed and then add the other ingredients. The beeswax may be melted with the neatsfoot oil and both added together, followed by the silicone. When the batch has been stirred to a smooth homogeneous mass, it may be run into cans to cool. The sodium silicate is used to prevent corrosion of the cans. The cans should be covered when the soap is cold, to avoid deposit of moisture on the lids.

P-S-615
Soap, surgical, cake and liquid

The specification covers surgical soap with hexachlorophene, suitable for use in preparing the operative field of the skin, and for scrubbing by surgical personnel before surgery.[2,3]

TYPE I: CAKE

Type I shall be a high grade, non-irritating thoroughly saponified sodium or sodium-potassium soap, in accordance with the following:

COMPOSITION

	% min	% max
Hexachlorophene USP	1.80	2.40
Moisture and matter volatile at 105°C	–	15.00
Sum of free alkali, total matter insoluble in alcohol, and NaCl	–	3.00
Free alkali, as NaOH	–	0.10
Matter insoluble in water	–	1.00
Unsaponified saponifiable matter	–	0.30
Anhydrous soap	81.00	–
Rosin	–	none
Sugar	–	none
Foreign matter	–	none

This should be a milled toilet soap, made by the whole-boiled process, with 20% coconut oil and 80% tallow.

The kettle soap is dried to chips and then the hexachlorophene is added during milling.

NOTE: See other bacteriostats, where hexachlorophene is not specifically ordered.

TYPE II: LIQUID

This shall be a clear liquid solution of potash soap in water. The soap shall be produced by the saponification of either vegetable oils or vegetable oil fatty acids with potash, in accordance with the following:

COMPOSITION

	% min	% max
Hexachlorophene	0.90	1.10
Anhydrous soap, as potash soap	18.00	22.0
Free alkali, as KOH	–	0.05
Free acid, as oleic	–	0.10
Alkaline salts as K_2CO_3	–	0.20
Matter insoluble in water	–	0.10
Alcohol-insoluble matter	–	0.50
Chloride, as KCl	–	0.30
Sodium and potassium oxides	–	0.50
Sulfates	–	trace
Rosin	–	none
Sugar	–	none
Iodine No. (Wijs) of fatty acids derived from the soap	7	80
Titer test of F.A. from soap	15°C	28°C
Acid No. of F.A. from soap	215	270

This soap should be made with coconut oil and saponified with potash, by procedure outlined under coconut liquid soap. The hexachlorophene may be added to the liquid while hot, stirred well until it dissolves; or a solution of the hexachlorophene in alcohol or propylene glycol may be made up and then added to the soap, either hot or cold.

P-S-00619 (GSA-FSS)
Interim Federal Specification (7/20/64)
Soap, Toilet, Cake, Anti-bacterial [11]

This specification covers soap containing anti-bacterial agents intended for personal use.

The material shall be nonirritating, thoroughly saponified, sodium or sodium-potassium soap conforming to the compositional requirements specified:

COMPOSITION

	% min	% max
Moisture and matter volatile at 105°C	–	15.0
Sum of free alkali, total matter insoluble in alcohol, and NaCl	–	3.0
Free alkali as NaOH	–	0.1
Matter insoluble in water	–	1.0
Unsaponified saponifiable matter	–	0.5
Anhydrous soap	79.0	–
Rosin		none
Sugar		none

The soap shall be milled colored or uncolored, well compressed in firm smooth cakes. It may be mildly perfumed.

Bacterial growth inhibition:

A 0.05% solution of the soap shall completely inhibit the growth of *Staphylococcus Aureus*, FDA 209 (ATCC No. 6538), when evaluated by the Agar Streak method. A 0.65% solution ±0.05%) of the soap shall completely inhibit the growth of *Escherichia Coli* (ATCC No. 11229) when evaluated by the Agar Streak method. These test methods are described in the specification.

Some of the germicides discussed under additives during milling of toilet soap may be used here, or as in preceding specification (P-S 615).

P-S-616d
Soap, Floating

This specification covers white floating soap, for use in personal bathing and general cleaning with soft water.

This should be a full-boiled soap, consisting of 25% of coconut oil and 75% of tallow. To attain an absolutely white soap, the coconut oil and the tallow should be bleached before saponification.

COMPOSITION

	Type I		Type II	
	% min	% max	% min	% max
Moisture and volatile at 105°C	–	34.0	–	24.0
Sum of free alkali, total matter insoluble in alcohol, and NaCl	–	2.0	–	2.5
Free alkali, as NaOH	–	0.1	–	0.1
Chloride, as NaCl	–	1.0	–	1.2
Water-insoluble matter	–	0.6	–	0.6
Anhydrous soap	62.0	–	73.5	–
Unsaponified saponifiable matter	–	0.3	–	0.3
Rosin (qualitative test)	–	none	–	none
Sugar	–	none	–	none
Acid No. of mixed fatty acids prepared from the soap	212		212	

Type I: Framed white floating soap, shall have a content of moisture and volatile matter, not in excess of 34 per cent.

Type II: White floating toilet soap, shall have a content of moisture and volatile, not in excess of 24% and shall be so processed as to be free from warping and distortion upon aging.

Unless this soap is made by a patent process [1] as described under production of floating soap with an even lower content of moisture, it will have to undergo considerable drying, before it is pressed and then require moisture-proof wrapping so as to retain its shape.

P-S-621e
Soap, Toilet (Cake-Milled)

The soap shall be thoroughly saponified and its composition shall conform to the following requirements:

COMPOSITION

	min %	max %
Moisture and volatile at 105°C	–	15.00
Sum of free alkali, total matter insoluble in alcohol, and NaCl	–	3.00
Free alkali, as NaOH	–	0.10
Matter insoluble in water	–	1.00
Unsaponified saponifiable matter (free fat)	–	0.30
Anhydrous soap	83.0	–
Rosin, sugar, and foreign matter	–	none

We have found, over a period of many years,that a full-boiled soap, consisting of 20% of Cochin coconut oil and 80% of fancy tallow, properly milled and pressed, gives satisfactory performance. A 200-lb milling batch should contain:

Soap flakes, about 12% moisture	200 lb
Pleasant perfume	1–2
Titanium dioxide*	0.5

* This is optional, to add whiteness, but it may be omitted if color is to be used.

There are many other ingredients that may be added in the milling, such as 0.5–1.0% of white mineral oil, lanolin, cholesterol or lecithin (as in Sweetheart soap); paraffin, petrolatum, beeswax, and dibromo- and tribromosaliculanilide (as in Cuticura soap USP 2 906 711); hexachlorophene and trichlorocarbanilide (as in Dial soap USP 2 535-077); tetrachlorosalicylanilide (Geigy's TCSA; as in Colgate's Coleo soap); trimethyl thiuramdisulfide (duPont); as in Life Buoy.

Swift's Lexard soap, besides being superfatted, contains hexachlorophene. Where hexachlorophene is unobtainable, Monsanto's Actamer (Bithionol) or Vancide BN (Vanderbilt) may be substituted.

P-S-624f
Soap, Toilet, Liquid and Paste

This specification covers liquid toilet soap ready for use in dispensers, and also a liquid and a paste soap which shall be diluted prior to use in dispensers.

TYPE I, LIQUID (15% minimum anhydrous soap)
TYPE II, PASTE (52% minimum anhydrous soap)
TYPE III, LIQUID (35% minimum anhydrous soap)

Odor

The soap shall be mildly perfumed and the odor shall not be objectionable as received, or in hot water. The soap shall not leave an objectionable odor on the skin after washing and rinsing with water.

Stability

A 15% (anhydrous soap content) solution of the material in distilled water, after having been frozen and then warmed to room temperature, shall exhibit no evidence of cloudiness or jelling.

Sudsing

The material shall produce a net volume of at least 100 ml of foam when tested as specified in 4.42.

Sequestering Agent
(Types II and III only)

The soap shall contain sufficient organic sequestering agent to give a clear solution when one part of Type II soap is diluted with three parts of water having a hardness of 500 ppm as calcium carbonate; or when 1 part of Type III soap is diluted with 1.5 parts of 500 ppm hard water, when tested as specified in 4.4.3.

***p*H value**

The *p*H value of the soap shall not exceed 10.5 when tested as specified in 4.4.4.

Colors

The liquid soap (Type I), as received, and the 15% solutions prepared from the paste soap (Type II) and liquid soap (Type III), shall not be darker in color than 10 on the Gardner Color Scale when tested in accordance with 4.4.5.

COMPOSITION OF LIQUID AND PASTE SOAPS

	Type I		Type II		Type III	
	% min	% max	% min	% max	% min	% max
Moisture (Toluene Distillation)	–	–	–	40.0	–	60.0
Anhydrous soap, as potash soap	15.0	–	52.0	–	35.0	–
Free alkali, as KOH	–	0.05	–	0.2	–	0.15
Free acid, as oleic	–	0.10	–	0.2	–	0.20
Alkaline salts, as K_2CO_3	–	0.20	–	0.2	–	0.30
Matter insoluble in water (distilled)	–	0.10	–	0.2	–	0.20
Matter insoluble in alcohol	–	0.50	–	1.0	–	1.0
Chloride, as KCl	–	0.30	–	0.5	–	0.50
Sodium compounds, as Na_2O	–	0.50	–	1.0	–	1.00
Sulfates	–	trace	–	trace	–	trace
Rosin	–	1.00	–	4.00	–	2.70
Sugar	–	none	–	none	–	none
Unsaponified plus unsaponifiable matter	–	0.60	–	2.40	–	1.60
Titer of mixed F.A. derived from the soap		32°C		32°C		32°C
Acid No. mixed F.A. derived from the soap	215	270	215	270	215	270

To meet the given specifications on the composition of the fatty acid, we would recommend the use of 25% of distilled tall oil, which does not contain over 10% of rosin and 75% of coconut oil. Such a composition when properly saponified with caustic potash, should yield the proper amount of suds, produce a soap of good color, and at the same time result in a product that will be less irritating to the skin, and much cheaper to make, than similar soap consisting entirely of coconut oil. As the titer of tall oil is low, the soap should have the proper fluidity to flow from a central reservoir to the dispensers. With such a small amount of tall oil the soap should be without offensive odor and can be pleasantly scented without excessive cost.

Procedure:

A soap base can be made and then the desired dilutions for Type I and Type III may be made from it, by adding water and the requisite amount of sequestering agent, color if desired and perfume.

Suggested Formula

To produce 100 pounds of the base Type II:

Distilled tall oil	12.5 lb
Coconut oil	37.5
Nullapon BF 78 (or Versene)	0.25
Potassium hydroxide (45%)	26.8
Water	22.95
	100.00 lb

Procedure:

The oils and about 30 pounds of water are placed in a jacketed kettle, heated to 180°F, and the potash is added little by little, allowing each portion to be absorbed before adding more. As the saponification proceeds, the mass will become slightly cloudy and it will also thicken until all the potash has combined, when it will become very translucent. If open steam coil is used in heating and at the same time to agitate the batch, it will not require the replacement of as much evaporated water (as we started with about 7% excess) but if the heating is conducted entirely by means of the jacket, then water must be added, so as to finish the soap, with the requisite moisture. Tests for neutrality and moisture should be made before withdrawing the soap.

When the various tests have been concluded and the soap found to be perfect, a solution of one-half pound of sequestrant dissolved in one quart of hot water, is added and boiled in. The soap may then be withdrawn.

TYPE I. The base soap described will contain between 52–55% of anhydrous soap, exclusive of glycerol. This is then diluted to a soap content of 15% by adding about 246 pounds of hot water, and heated while agitating, until the entire volume is liquid. It may then be tested as before. When cold, it may be filtered through Sorbo-cel (Johns-Manville) or similar type of filter medium. An alternate method is to allow the soap to cool and settle for several days, when a very clear liquid soap may be obtained. For best results the liquid should be cooled to 40°F and then filtered, using Sorbo-cel or similar amporphous silica. A filter press will facilitate the filtration.

TYPE III. To obtain the required concentration, about 50 pounds of hot water will be required for the 100 pounds of soap base in the kettle. It is rather difficult to obtain a uniform mix after the base soap has stopped boiling, so for Type III it should be made from the start with the amount of water that will be required when finished.

If the soaps are to be colored and/or perfumed, especially Type II, the additions should be made at the end of the saponification. For opalescent coloring, which is popular for liquid soap, D&C Yellow #8 is recommended.

Alternate Product

An alternate soap consisting wholly of coconut oil may be produced in a similar manner, using the proper amount of potash for the saponification. 50 pounds of coconut oil, will require:

$$\frac{50 \times 0.260}{0.45} = 28.1 \text{ lbs of } 45\% \text{ KOH}$$

The batch now will consist of:

Coconut oil	50 lb
Water	21.65
Nullapon BF 78	0.25
Potash (45%)	28.1

The procedure will be similar to that described and it is suggested that some extra water be added at the start.

P-S-628a

Soap, Borax, for Dispensers

TYPE I: Without lanolin
TYPE II: With lanolin

COMPOSITION

	TYPE I		TYPE II	
	% min	% max	% min	% max
Anhydrous soap	23.0	27.0	35.0	39.0
Borax, as $Na_2B_4O_7 \cdot 10H_2O$	72.0	77.0	57.0	63.0
Lanolin	–	0.0	2.5	3.5
*p*H value	9	10.2	8.5	9.5
Matter insoluble in water	–	0.2	–	0.3

For the basic soap, we would recommend granulated toilet soap (7% moisture) such as P-S-621e.

To compound Type I, mix 50 pounds of soap with 150 pounds borax in the amalgamator or in a ribbon-type mixer. If desired it may be suitably perfumed.

For Type II, mix 80 pounds of soap (containing about 7% of moisture) with 115 pounds of borax, and while mixing pour over the batch, from one end of the mixer to the other, a solution of 5 pounds of anhydrous lanolin, dissolved in one gallon of deodorized kerosine, or in 10 pounds of Chlorethene (Dow). After mixing, the batch should be spread out on the floor, until the solvent has evaporated. It may then be packaged. Here, too, a suitable perfume can be added during mixing.

FFF-C-00641 (GSA-FSS)
Cream, Shaving and Soap, Shaving
(Interim Specification)

TYPE I: CAKE SOAP
TYPE II: SHAVING CREAM
Class 1: Lather
Class 2: Brushless
TYPE III: SHAVING CREAM in pressurized dispenser.

Type I shaving soap shall be furnished in a 2-ounce cake. Odor shall be mild and pleasant and color of a freshly cut surface shall be not darker than No. 37855 of Federal Standard No. 595.

COMPOSITION

	% min	% max
Matter volatile at 105°C	–	10.0
Matter insoluble in alcohol	–	0.8
Free alkali, as KOH	–	none
Free fatty acids, as stearic	–	1.0
Matter insoluble in water	–	0.4
Anhydrous soap, as potash soap	87.0	–

Suggested Formula

To make 100 pounds, the following will be required:

Cochin coconut oil	14.5 lb
Fancy tallow	29.0
Stearic acid	29.0
Caustic potash (45%)	17.6
Caustic soda (50%)	10.5
	100.0 lb

This soap is produced by the half-boiled process.

Procedure:

The fats are heated to 180°F in a jacketed kettle or, if the batch is sufficiently large, in the crutcher. About 2 pounds of the stearic acid may be withheld, as the finished soap requires 1% content of free fatty acid calculated as stearic. The mixed alkalies are added slowly, little by little, while the batch is being stirred, until the saponification is completed. The soap should now be tested for free alkali or free fatty acid, and proper adjustment made by adding the 2 pounds of stearic acid previously withheld. The finished product must show a content of 1% of free fatty acids. It should also be tested for the moisture. When the soap has cooled somewhat, perfume may be added and the soap may be withdrawn for cooling, chipping, milling, plodding, and pressing. In other words, it may be handled as described under milled toilet soap.

For the manufacture of this type of soap, nickel-clad kettles are used, so as to preclude any possibility of the soap becoming discolored by iron rust.

TYPE II SHAVING CREAM shall be a homogeneous product having a mild and pleasant odor, furnished in a collapsible tube of the size specified.

COMPOSITION

	Class 1		Class 2	
	% min	% max	% min	% max
Moisture (toluene distillation)	–	50.0	–	70.0
Matter insoluble in alcohol	–	0.3	–	–
Free alkali	–	none	–	none
Free fatty acid, as stearic acid	–	6.5	15.0	–
Matter insoluble in water	–	0.3	–	–
Anhydrous soap, as potash soap	40.0	–	–	–

Class 1: Lather Cream

Stearic acid	35.00 lb
Cochin coconut oil	8.90
Glycerol or propylene glycol	5.00
Potash (45%)	17.00
Borax	0.70
Water	33.05
Menthol	0.175
Lavender oil	0.175
	100.00 lb

Procedure:

The stearic acid, coconut oil, glycerol, and water are heated in a jacketed kettle or crutcher, to 175°F. The potash is added slowly, while stirring, until saponification has taken place. At this point, a test should show that there is about 5% of free fatty acids; otherwise, the soap should be adjusted accordingly and the entire batch should be mixed until uniform. The borax is dissolved in a small amount of hot water and added to the mixture. The soap is now allowed to cool, after which it is scented by a solution of the menthol crystals in the lavender oil. In a day or so, the soap will turn pearly. It should be packed at room temperature.

Class 2: Brushless Shaving Cream

Stearic acid		16.0 lb
Paraffin		2.0
Sulfonated castor oil		1.0
Cottonseed oil		2.0
Potash (45%)		1.1
Water		74.1
Glycerol		3.0
Perfume:		
Phenol	0.05	0.8
Eucalyptol	0.25	
Menthol	0.25	
Lavender oil	0.25	
Total		100.0 lb

Procedure:

The method followed is practically the same for the preceding. A test should be taken for moisture and for free fatty acids, to ascertain whether the product conforms to specification. The mixed perfume is stirred in, before filling.

TYPE III SHAVING CREAM

The shaving cream in pressurized dispenser shall consist of a soap-propellent mixture furnished in a dispenser of the capacity specified.

The soap component shall be not less than 5% nor more than 12 per cent.

The propellent part shall be not less than 2.4% of the mixture, if pressurized with hydrocarbons, or not less than 5% if pressurized with fluorinated hydrocarbons. The hydrocarbon propellent shall be composed of a mixture of isobutane and propane. The fluorinated hydrocarbons may be dichlorodifluoromethane (Propellent 12) or a mixture of dichlorodifluoromethane (12) and dichlorotetrafluoromethane (Propellent 114).

To produce a pressurized product from the lathering cream (Type II, Class one) the material must be diluted with water, at least one part of the cream to three parts of hot water. This is made up into a uniform solution before it is loaded into the aerosol cans. Trial samples should be made up and tested for consistency, stability, etc.

Class 2, brushless, must also be diluted to about the same extent and subjected to all the tests, as for Class 1.

Having once ascertained the exact dilution required, later batches should be made accordingly.

The stability of the lather should be tested by ejecting the lather from each type onto a glass surface, to note whether the lather remains or falls flat. Commercial products should be used, for comparison. Actual shaving tests should also be made.

REFERENCES

1. (Floating Soap) USP 2 210 924; 2 215 539 (1941).
2. USP 2 159 397.
3. USP 1 968 526; 2 249 676.
4. Some patents on spray-drying of soap and other products: USP 271 281; 890 078; 940 398; 1 484 271; 1 551 557; 1 472 473; 1 603 559; 1 621 506; Renewal 16 749; 1 718 778; Renewal 18 546; 1 732 454; 1 734 260; Renewal 19 556; 1 785 533; 1 804 932; 1 887 474; 1 897 701; 1 971 566; 2 043 378; 2 046 449; 2 123 647; 2 152 788; 2 159 397.
5. F. E. Wall, *Principles and Practice of Beauty Culture* (4th Ed.); Keystone Publications, Inc., New York, 1960.
6. R. D. Stayner, *Mechanism of Cotton Detergency*; Oronite Division, Chevron Chemical Co., San Francisco, Calif.
7. B. Levitt, Oxidation of Soap and Its Practical Prevention, *Indian Soap Journal*, March 1936.
8. DDT in Soaps, *Nature*, 158:156 (1945).
9. P. Weggeman, Brighteners in Detergent Specialties, *Soap Chem. Specialties*, December 1965.
10. M. Fishbein, *Medical Uses of Soap*; Lippincott, Philadelphia, 1945.
11. A. R. Cade, An *In vivo* Method of Determining the Degerming Activity of Soaps Containing Hexachlorophene; *Special Technical Publication No. 15*; American Society for Testing and Materials, 1952.

chapter 8

SYNTHETIC DETERGENTS

For small manufacturers, the best advice on making synthetic detergents is to purchase from primary producers or their representatives, the exact type of surfactant required for a specific purpose. On the other hand, formulators of substantial quantities of either liquid or solid detergents, can avail themselves of the economy in purchasing an intermediate dodecylbenzene sulfonic acid (DDBSA) or Linear Alkyl Sulfonic acids. They, themselves, can neutralize the dodecyl compound in a variety of ways. They may use ammonia, amines, soda, or potash to yield solids or liquids, all from the same intermediate.

DDBSA is easily incorporated into dry formulations and makes possible great variations because of its low water content and high active content. Homogeneous dry products with less tendency to segregate are produced. No investment in drying equipment is required. The physical form of the product can be varied, depending on the formulation, from powder to a granular material.

TABLE 8:1

PERCENTAGE COMPOSITION OF DODECYLBENZENESULFONIC ACID

Supplier	*Active DDBSA*	*Sulfuric Acid*	*Free Oil*	*Water*	*Mol. Wt. DDBSA*	*F.O.B. Point*
Continental Oil Co.	88.0	7.0	1.2	3.8	323	Chicago, Ill.
Continental Oil Co.	99.5					Hammond, Ind.
Monsanto	88.5	8.7	1.7	1.1	320	E. St. Louis Eq. w. Chicago
Nat. Aniline (Allied Chem.)	88.7	6.6	1.9	2.8	321	Buffalo, N.Y.
Oronite	84.0 min.	10–15	2.0 max.			Richmond, Calif.
Pilot, Calif.	98.0	1.0	1.0	nil		Santa Fé Spgs., Calif.

DDBSA is an opaque, dark brown viscous liquid at room temperatures of 20–25°C. The molecular weight of the active portion is approximately 320.

Typical analyses of dodecylbenzenesulfonic acid from five different producers are shown in Table 8:1.

Weight Equivalents of DDBSA

The weight equivalency of DDBSA and active dodecylbenzenesulfonic salts is as follows:

MOLECULAR WEIGHTS

DDBSA (Active)	321
NaOH	40.01
Na_2CO_3	106
NH_4OH	35.05
Triethanolamine	149.19
Dodecylbenzenesulfonate:	
Sodium salt	343
Ammonium salt	337
Triethanolamine salt	451

One pound of DDBSA (88% active), when neutralized, will yield on an active basis:

0.94 pound of sodium dodecylbenzenesulfonate
0.92 ,, ,, ammonium ,,
1.24 ,, ,, triethanolamine ,,

or a quantity of active, equal to 2.32 pounds of Santomerse No. 1 (which contains 40% active).

To make one pound of dodecylbenzenesulfonic salt, the amount of DDBSA (88% active) required is:

1.06 pounds for the sodium salt
1.09 ,, ,, ,, ammonium
0.81 ,, ,, ,, triethanolamine salt

Special Procedures for Compounding

Following, is an example of the procedure for compounding a 100-gal batch (about 870 lb) of liquid detergent, which is to contain 35% of active surfactants, of which 23% will be sodium alkylaryl sulfonate and 12% of Triton X-102; the active content will be comprised of 66% of sodium alkylaryl sulfonate and 34% of Triton X-102.

The batch will actually contain 870 × 0.23 or 200 pounds of the sulfonate and 870 × 0.12 or 104 pounds of Triton X-102.

To prepare 200 pounds of anionic will require 200/343, or 0.58 moles, equal to 187 pounds of 100% sulfonic acid. This is equivalent to 212 pounds of the sulfonic acid as supplied.

The 212 pounds of sulfonic acid contain 13.9 pounds or 0.14 moles, of free sulfuric acid, which is to be neutralized by 0.28 moles or 11.2 pounds of sodium hydroxide. The amount of sodium hydroxide required to neutralize the sulfonic acid is 0.58 moles or 23.2 pounds; therefore, the total anhydrous caustic soda to neutralize both the free sulfuric and the sulfonic acids is 34.4 pounds.

To summarize the calculation thus far, the batch requires:

DDBSA (National Aniline)	212.0 lb
Triton X-102	104.0
NaOH	34.4
Water	519.6
For total weight	870.0 lb

The formulator may wish to include a small amount of supplemental ethanol or isopropanol, to ensure homogeneity and control viscosity. This formulation also lacks lauric diethanolamide, which may be included to boost suds and improve viscosity. The latter is optional, as conditions may dictate.

The kettle, which should be either glass-lined or of suitable construction in stainless steel, is charged as follows:

Water	400 lb
Anhydrous caustic soda	34.4
Triton X-102	104.0
Sulfonic acid	211.0

Water is added to the caustic soda and stirred gently until dissolved; or a 50% solution of liquid caustic may be used (about 69 lb). The Triton X-102 is added while stirring.

The sulfonic acid is added slowly, with mild, sub-surface agitation, so that no air is entrained. A temperature rise of some 65°F (36°C) may be expected, so that cooling of the reaction vessel may be considered. If the color development (darkening) is not severe, cooling may not be necessary.

During this step, the rest of the water (120.6 pounds) is added, if 100% caustic has been used; if 50% caustic was used, then only 86 lbs of water should be added. The water will also help to control the temperature. The mix will probably be slightly acidic at this point, requiring an addition of caustic soda, to correct to the desired 7 *p*H. The addition should be made in the form of 25% NaOH solution, little by little, with stirring, until the final end point is reached.

When the mix is entirely cool, add about 3% (26 pounds) of alcohol, along with perfume, color, etc., if desired.

Other compositions can be produced similarly for various cleaning purposes. In each case, however, the calculations will change somewhat, depending on the composition of the starting acid and the final product desired.

Suggested Formulas

As powdered detergents are low-foaming and inexpensive, Wyandotte suggests the following

Plurafac AC-26	10.0%
Carbose D (carboxymethylcellulose)	1.5
Sodium metasilicate (penta)	10–15
Sodium tripolyphosphate, granular	50–65

Rohm & Haas suggests the following as low-foam efficient detergents:

	A	*B*
Triton X-114	5%	5%
Sodium tripolyphosphate	40	40
Sodium sulfate	30	–
Sodium silicate GD* or equal	10	15
Borax	14	–
Soda Ash	–	39
CMC	1	1
Water-soluble optical bleach	0.003	0.003

* Supplier: 114.

Rohm & Haas made a study of five commercially available alkylaryl sulfonic acids, to learn how they might best be formulated into liquid detergents, using Triton X-102 as a companion surfactant.

A major contribution of Triton X-102, in addition to foam stability and detergency, is its solublizing or hydrotropic effect upon the sodium alkylaryl sulfonate. This effect varies with the particular sulfonic acid used. Sometimes, a specific amount of supplemental hydrotrope (i.e., alcohol) may be required.

In this testing program, samples were rated on their appearance after being exposed to three freeze-thaw cycles (18 hours at −10°C, followed by 18 hours at 25°C, with no agitation). To be termed satisfactory, the sample must show no evidence of separation or stratification. It must be completely homogeneous throughout.

TABLE 8:2

HOMOGENEITY OF LIQUID DETERGENTS*

Source of Sulfonic Acid		*Active Content of Liquid Product*		*% Isopropanol for Homogeneity*	*Physical Condition after Three Freeze-Thaw Cycles*
		Total %	% NaAAS/ % Triton X-102		
Continental Oil Co.	a)	40	70:30	0–10	Not satisfactory; sep'n or salting out
	b)	35	70:30	4–6	Satisfactory
	c)	40	60:40	3	Satisfactory
		35	60:40	3–5	Satisfactory
Monsanto Chemical Co.		40	70:30	5	Satisfactory
		35	70:30	5	Satisfactory
		40	60:40	3–4	Satisfactory
		35	60:40	3–4	Satisfactory
National Aniline Div. Allied Chemical Corp.		40	70:30	2–5	Satisfactory
		35	70:30	2–4	Satisfactory
		40	60:40	0–2	Satisfactory
		35	60:40	0–5	Satisfactory
Oronite Chemical Co.		40	70:30	0–10	Not satisfactory; sep'n or salting out
		35	70:30	5–6	Satisfactory
		40	60:40	4	Satisfactory
		35	60:40	3–6	Satisfactory
Pilot California Co.		40	70:30	5–6	Satisfactory
		35	70:30	4–6	Satisfactory
		40	60:40	3	Satisfactory
		35	60:40	0–3	Satisfactory

* Prepared by Rohm & Haas Co.

Table 8:2 shows the homogeneous characteristics of liquid blends of sulfonic acids from the major producers, with Triton X-102, in detergents containing either 35 or 40% of total active surfactants.

As examples, here are three powdered detergents, low-foaming and inexpensive:

Wyandotte suggests this, which includes one of their biodegradable Plurafacs:

Plurafac-AC-26	10.0%
Carbose D (CMC)	1.5
Sodium metasilicate (penta)	10–15
Sodium tripolyphosphate (granular)	50–65

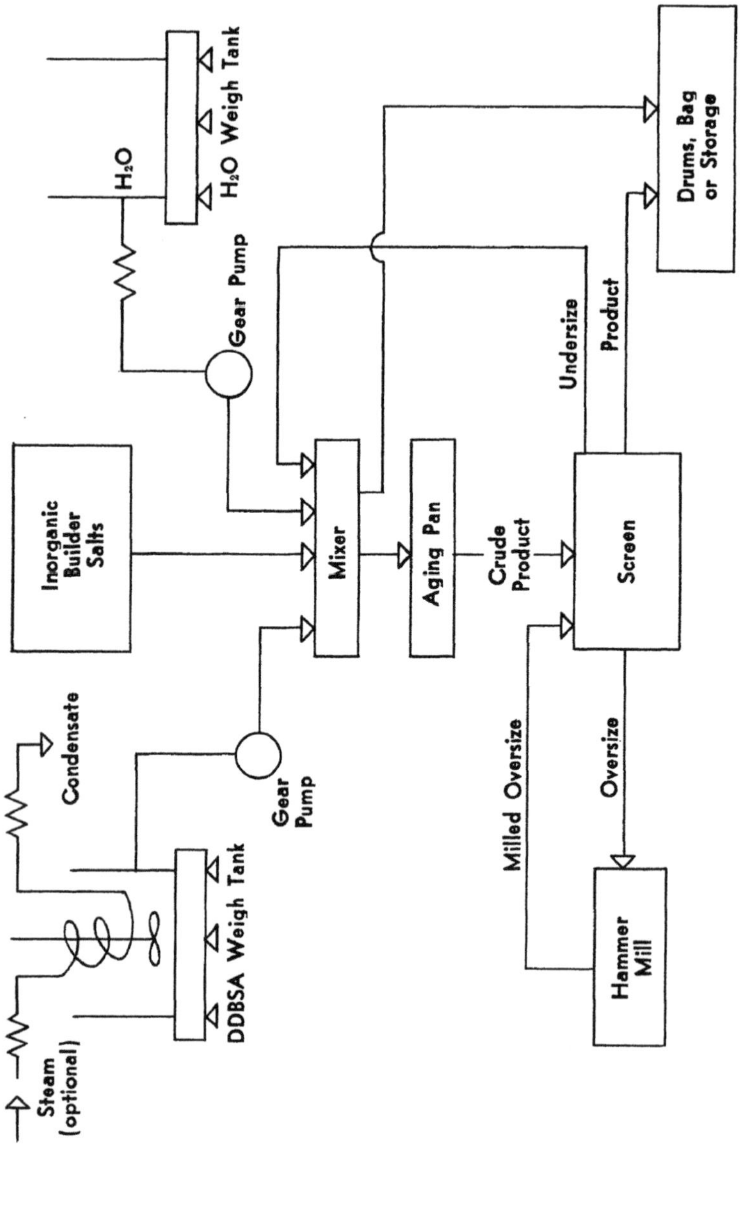

FIG. 8:1. FLOW-SHEET FOR DRY NEUTRALIZATION OF DDBSA
(Monsanto Company, St. Louis, Mo.)

Rohm & Haas suggests these, as low-foam efficient detergents:

	A	*B*
Triton X-114	5%	5%
Sodium tripolyphosphate	40	40
Sodium sulfate	30	–
Borax	14	–
Sodium Silicate* GD or equal	10	15
Soda Ash	–	39
CMC	1	1
Water soluble optical bleach	0.003	0.003

* Supplier: 114.

A word of caution may not be amiss. Small experimental batches should be tested for physical properties, as outlined in the study made by Rohm & Haas. They should also be tested for several months for stability in shelf life before a product is finally marketed.

The following are other examples of a light-duty liquid detergent:

(1)

DDBSA	18.5%
30% Sodium Lauryl Sulfate	36.7
Lauric Diethanolamide	10.5
26° Aqua Ammonia	5.0
Alcohol	15.0
Water	14.3
	100.0%

This will contain approximately 35–38% active and will show a cloud point of about 14°F.

(2)

DDBSA	20.0%
Nonionic Sterox AJ*	23.5
Water	49.1
Caustic Soda, 50%	7.4
	100.0%

* Supplier: 93

The viscosity may be reduced by adding 5% of alcohol.

DRY DETERGENTS

Dry or granular detergent may be produced, according to Monsanto Chemical Co., from DDBSA, by neutralizing with sodium carbonate. A variety of particle sizes may be controlled by the choice of builders and the quantity of water added to the formulation, in addition to proper speeds and conditions of mixing. This should be determined by making small experimental trials.

Dry Neutralization

The dry neutralization technique involves the addition of DDBSA to soda ash and/or alkaline builders, if any, already in the mixer. The ribbon-type of mixer has been found desirable to give bead-type particles, but other mixers are equally applicable if they can be adapted for the addition of the DDBSA while the mixer is in motion. A recommended flow sheet for neutralization of DDBSA is shown in Figure 8:1.

To make 1000 lbs of a 40%-active alkylaryl sulfonate:

Monsanto's 88% active DDBSA is equivalent to 0.94 pound of the sodium salt. Therefore to obtain 400 pounds of the sodium salt, we shall require 425.5 pounds of DDBSA. This weight of DDBSA is equivalent to 374 pounds of 100% DDBSA.

It must be remembered that 425.5 pounds of the sulfonic acid contains 8.7% of sulfuric acid, which must also be neutralized.

$\frac{374}{321}$ equals 1.16 moles of DDBSA to be neutralized by 53×1.16 moles or 61.5 pounds of soda ash.

$\frac{37}{98}$ equals 0.38 moles of sulfuric acid to be neutralized. 106×0.38 equals 40 pounds of soda ash. The total sodium carbonate necessary to neutralize both acids, therefore, is 101.5 pounds.

In making this compound, the inorganic salts, including inert material such as sodium sulfate, are added to the mixer, which is then started in motion. Next, the calculated amount of DDBSA is poured in from a bucket, through spray nozzles or by means of a perforated pipe, over the mixer. Following that, water (about 3% of the total weight of the batch) is added to initiate the reaction of neutralization.

The product may be dumped into aging pans or trays; this usually requires only a few minutes. It is either packed as is, or screened and then bagged, with undersize being returned to the mixer, and oversize if it is a significant quantity, being ground by a low-speed hammer mill and returned to the screen.

If a powdered rather than a beaded product is required, it is usually desirable to add enough soda ash or sufficient of the other ingredients, to absorb the DDBSA that is added, so that the remaining ingredients will mix in easily.

Spray nozzles should be of a type to give a flat spray, and should be made of 316 stainless steel. As DDBSA is rather viscous, it is necessary to preheat it for spraying. A temperature of 55°C should be attained for best use of spray jets.[1]

The temperature of the mixer will rise when DDBSA is neutralized by alkaline salts, often as high as 150°F in large batches. The calculated heat of neutralization is approximately 200 BTU to the pound at 20°C.

When finished, if the compound is still acidic, more soda ash may be added to bring the mixture to the neutral point. If it should be too alkaline, it may be brought down to *p*H 7–8 by the addition of dilute phosphoric acid or acid phosphate.

For spray-drying, in order to obtain a bulky, light-weight bead, it would be necessary to include considerably more water, to facilitate the flow and for evaporation during the drying. Many examples of various detergents for commercial and household use, embodying some of the surfactants mentioned are cited throughout the book and in Volume 2. Where abstracts of patents are cited, it is not with any intention of sanctioning infringement, but merely to give examples of what may be found on the American market. With a little ingenuity on the part of a compounder, it is quite possible to produce commodities of equal worth (Fig. 8:2).

Tests for moisture on retail products may be used as guides for compounding products for competitive sale (see chapter 9).

Detergent Tablets

U S. patent 3 154 494 by Richard C. Speak and P. H. MacConnell, assigned to U.S. Borax & Chemical Co., describes a laundry tablet made with an inner core of dry bleach, surrounded by a coating that may contain detergent, water conditioners, optical brighteners, bacteriostats, bactericides and perfume. The tablet releases the bleach one to 5 minutes after the detergent mixture has dissolved. It is explained that delaying the addition of bleach in the laundry operation offers the advantage of non-interference with the cleaning action of certain detergents. It also prevents any untoward reaction of the solid bleach, with the other components, that might reduce its potency, while in storage.

USES OF SULFONIC ACIDS

In the manufacture of detergents, a sulfonic acid is first neutralized with an appropriate base, such as NaOH,NH_4OH,KOH or an amine,

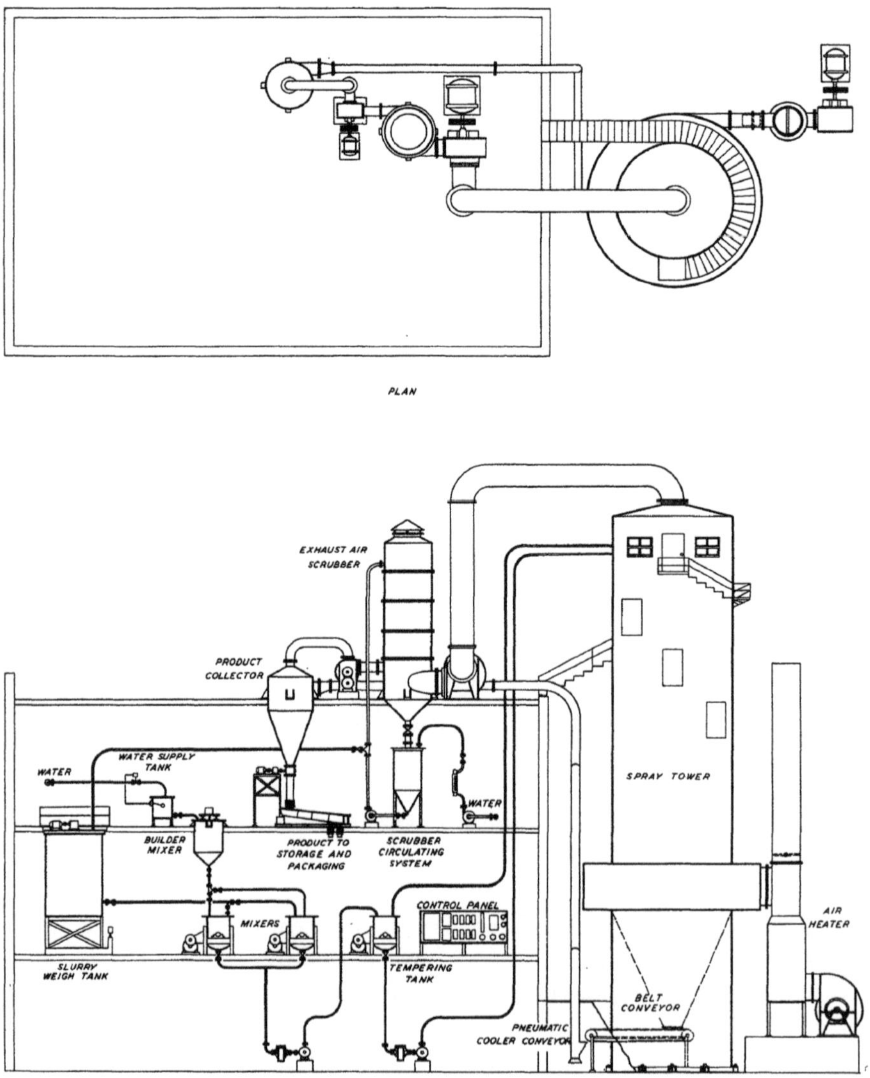

FIG. 8:2. CONTINUOUS SPRAY PROCESS IN SYNTHETIC DETERGENT PLANT
(*Wurster & Sanger, Inc., Chicago, Ill.*)

to produce a sulfonate. The product may be used for liquid or dry formulations.

Sulfonation of Alkyl Benzene

A continuous process for sulfonation of alkyl benzene has been patented.[2] It is rapid, with continuous neutralization which is provided by means of a series of centrifugal pumps and heat-exchangers. The equipment is small but by rapid pumping and recycling, quantity production is attained with completeness, leaving only 0.2% of unreacted alkylbenzene, in a 40%-active detergent. By injecting the components at room temperature through a concentric pipe, into the first pump, running at 1750 rpm, the alkyl benzene is injected at 3.5 pounds a minute, and the 22% fuming sulfuric acid is run at 4.21 pounds a minute. The exothermic reaction is maintained at 140°F maximum.

Simultaneously the partly reacted solution is recycled into the pump to absorb some of the heat. The mix then passes to the first heat-exchanger, where the temperature is reduced to 120°F. The mix is split, one part is circulated through the first pump, the second goes to the reaction chamber. The rate of flow of the recycled portion compared to the volume of the entering reactants, is approximately 15:1. Four to ten minutes is the reaction time. The time in the mixing pump does not exceed three minutes. The addition of 0.77 pounds of water a minute, stops the reaction.

The reaction mix and water are then transferred by a second pump to another heat-exchanger where the solution is again split. Part is circulated by the second pump along with fresh reaction mix and water, and the second portion runs into a separating tank, where the water and sulfonic acid separate. The sulfonic acid is mixed in a third pump with 14.5% of caustic soda solution, which yields a slurry of *p*H 10–10.5. A heat-exchanger reduces the temperature to 110°F. The slurry is split into two portions, one circulating to the pump along with the caustic solution and fresh sulfonic acid, the second, is further neutralized to yield a detergent with a *p*H of 7 and composition as follows:

Sodium alkylsulfonate	40.0%
Unreacted alkylbenzene	0.2
Sodium sulfate	5.0
Water	54.8

A flow sheet for this process is shown in Figure 8:3.

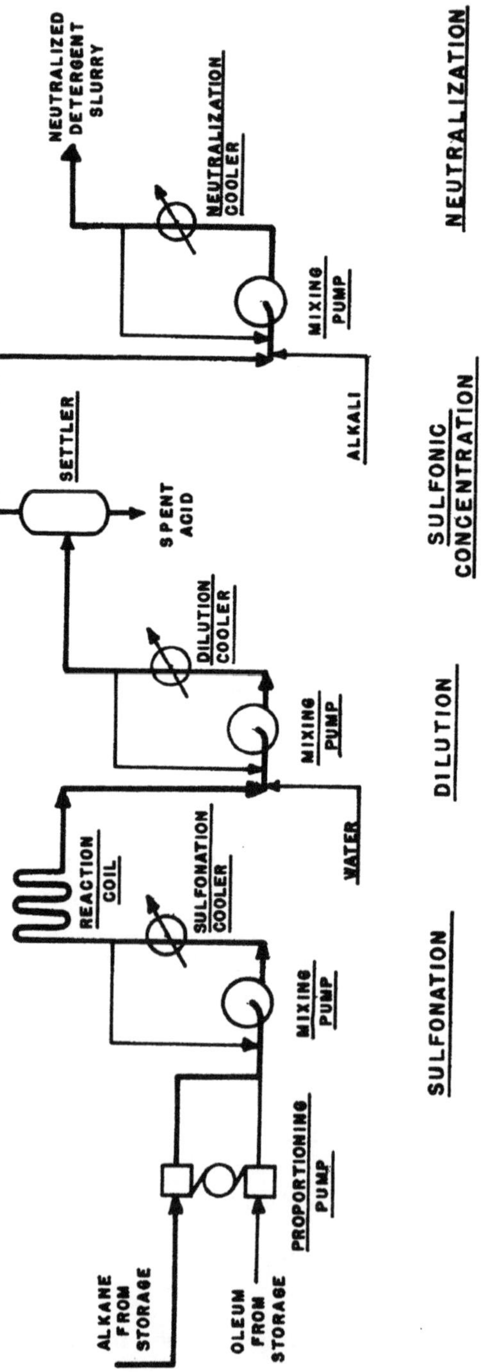

FIG. 8:3. FLOW-SHEET FOR CONTINUOUS HIGH-ALKANE SULFONATION
(The Chemithon Corp., Seattle, Wash.)

BIODEGRADABLE DETERGENTS

The Petrochemical Department of Continental Oil Company offers Nalkylene-based, linear alkylate sulfonate (LAS) materials. It is claimed that LAS slurries made from Nalkylene 500 and Nalkylene 600, offer full biodegradability, for all types of formulations, conventionally using ABS (alkylbenzene sulfonates) hard detergents.

FIG. 8:4. TYPICAL CHEMITHON PLANT
(*The Chemithon Corp., Seattle, Wash.*)

The Nalkylene detergent intermediates are derived from straight-chain paraffins and benzene. They are blends of linear alkylates. Nalkylene 500 shows average hydrocarbon length of approximately C_{12} and substitutes for hard dodecylbenzene. Nalkylene 600 substitutes for hard tridecylbenzene.

The acids of lower molecular weight derive from Nalkylene 500 through sulfonation, with oleum for the SA 588 product and with SO_3 for the SA 597. Nalkylene 600 yields SA 688 and SA 697, through the same procedures with oleum and SO_3.

The linear acids show much lower viscosity than their branched-chain counterparts.

A maker of conventional hard detergents can substitute his branched-chain alkylbenzenesulfonate with a Nalkylene sulfonate, frequently at lower active content, for high-performance heavy-duty products. Reformulating without foam stabilizers can yield equal or better results in dishwashing, at equal or lower cost in light-duty compounds.

TABLE 8:3

TYPICAL PHYSICAL PROPERTIES OF BIODEGRADABLE SULFONIC ACIDS

	Conoco SA 588	*Conoco SA 597*	*Conoco SA 688*	*Conoco SA 696*
Av. Mol. Wt.	318	318	341	341
Act. Ingr. Wt. %	88.9	97.1	88.9	97.2
H_2SO_4 " "	7.3	1.1	7.3	1.0
Free Oil " "	1.0	1.1	0.9	1.2
Water " "	2.8	0.7	2.9	0.6

Nalkylene-based sulfonic acids serve in the manufacture of a broad range of biodegradable surfactant products, which include liquid and dry detergents for light and heavy duty products; toiletries (shampoo and bubble-bath compounds); industrial cleaners (car-wash, dry cleaning aids); industrial detergents (emulsifiers, cutting oils, oil sprays); Dispersants (ore flotation surfactants); polymerization stabilizers, chemical intermediates, sulfonamids and esters.

TABLE 8:4

PROPERTIES OF NALKYLENE LAS COMPOUNDS

	Conoco C 550	*Conoco C 560*	*Conoco C 650*	*Conoco C 660*
Mol. Wt.	341	341	364	364
Act. Ingr. %	45.0	60.0	45.0	60.0
Na_2SO_4	3.9	1.2	3.9	1.1
Free Oil %	1.1	0.8	1.0	0.8
Water %	50.0	38.0	50.1	38.1
*p*H @ 25°C	7.5	7.5	7.5	7.5
Color Klett*	55.0	60.0	55.0	60.0

* 4% active solution, 4 cm cell, #4 filter.

The line includes four LAS slurries of straight-chain sodium alkylbenzenesulfonate, as shown in Table 8:4.

The Conoco soft slurries of lower viscosity provide easier pumping and mixing. Also, the conventional branched-chain materials show a rapid increase in viscosity above 150°F, not observed in the linear slurries described.

Conoco C 550 can be used for making industrial and household solids. Conoco C 560 slurry is primarily for liquid products. It requires less hydrotrope than the counterpart ABS slurry. As a rule, in aqueous systems, the higher cloud points occur regularly at higher active levels. The linear products show lower cloud points than do the ABS products at equal concentrations. The cloud points in aqueous solution rise faster at the higher active levels as the ratio of Na_2SO_4 to active ingredient increases.

For low (10–15% active) solutions, the content of sodium salt (about 3% of the active for SO_3-sulfonation, and 9% for oleum) has little effect in either aqueous or alcoholic formulations. The cloud point drops as the content of ethanol rises.

Interfacial tension of Nalkylene sulfonates indicates definite advantage for general-purpose cleaners. Improved solubility along with lower interfacial tension promotes improved cleaning and dispersion where a hydrocarbon or pine oil is used in the product, for removing grease and oil from floors, walls, work clothes, and other badly soiled surfaces. Especially is this noticeable in the low interfacial values obtained with isopropanolamine sulfonates. The surface tension of Nalkylene sulfonates is lower than that of their ABS counterparts. In testing ABS and LAS products as wetting agents for applications in dust-control, Nalkylene-based LAS will wet solid powders better, at lower concentration than will branched-chain ABS products.

SPECIFICATIONS FOR SYNTHETIC PRODUCTS

P–D 240 Detergent, Surgical

An anti-bacterial non-alkaline, non-irritating emollient, sudsing, liquid cream detergent, suitable for rapid reduction of skin surface bacteria, by washing and scrubbing. A stable homogeneous white-to-cream-colored preparation without added coloring, consisting of:[3]

Hexachlorophene USP	3% W/W
Detergent	50
Emollients	8
Inert vehicle q.s.	100%

Detergent agent—Alkylaryl polyether sulfonate, containing 28% ± 3% solids.
Emollients shall be lanolin cholesterols and petrolatum.

*p*H 5–6
Visc 5–20 seconds at 25°C
Total solids 29–34%
Phenol coefficient, 1.5
Size: 5 oz to 1 gal

The following formulation is suggested as typical in many respects of the requirements listed and also is somewhat similar to the well known product, pHisohex covered by US Patent 2 303 932:

Triton X 200	45%
,, X 15	2
Sodium Lauryl sulfate (30%)	45
Hexachlorophene	3
Liquid petrolatum	1
Propylene glycol	2
Amerchol L 101	2
Perfume—optional	
	100%

Melt together the first four ingredients and add the others in sequence. Stir until a uniform emulsion is obtained.

P-S-617

Soap, Toilet, Hard, Soft, or Sea Water

This specification covers a toilet soap bar for use in fresh- or sea water bathing, shaving, and hand laundering.[4]

The finished toilet soap bar shall be composed of a compatible mixture of synthetic organic detergents, foaming agents, sodium carboxymethylcellulose, sodium silicate, and mild perfume, with or without high quality sodium or potassium soap.

Consistency:

The soap bar shall be firm and homogeneous in appearance corresponding to commercially acceptable toilet soap and shall be non-crumbling and non-abrasive, either initially or after intermittent use. A sample toilet soap bar when tested according to the method prescribed, shall not show any cracking or tendency toward crumbling and the surface shall be smooth and nonabrasive.

COMPOSITION

	% min	% max
Moisture and volatile matter at 105°C	–	12.0
Anhydrous salt-free synthetic organic detergent	25.0	–
Sum of anhydrous sodium or potassium soap and synthetic organic detergent	65.0	–
Free alkali, as NaOH	–	0.2
Free acid, as oleic	–	0.2
Rosin	–	none
Sodium silicate as $Na_2O:3.25\ SiO_2$	0.5	2.0
Sodium carboxymethyl cellulose (active ingredient basis)	1.0	–
Water-insoluble matter	–	2.0
Starch	–	none

The *p*H of a 0.5% solution of the soap in distilled water shall be 8.5–10.3 when tested as specified.

Lather:
The soap bar when used with synthetic sea water in washing the hands in the customary fashion, shall have a soapy "feel" and shall produce a lather when tested as specified.

Sudsing in sea water:
A 1% solution of the soap bar in synthetic sea water shall yield a minimum of 100 ml. of suds when tested as specified.

Soil removal from skin:
The soap shall remove grease from the skin in one washing with synthetic sea water at 20°–25°C when tested as specified.

Workmanship:
The soap shall be a well-cut milled bar of the specified weight and shape and shall conform to the quality specified.

A close approximation to these specifications, suggested by Antara Chemicals, is as follows:

Tallow Soap	54.7%
Igepon AC-78	35.3 (to give 30% active)
CMC	2.0
Sodium metasilicate	1.0
Water	7.0
	100.0%

Another formulation which has been worked out to match the above specification is:

Soap chips	50%
Nacconol NRSF	15
Sodium lauryl sulfate (90%)	20
CMC (100% basis)	1
R-Brand silicate	4
Water	10
	100%

The ingredients may be mixed in an amalgamator and then milled in the conventional way. At a moisture content of 13%, the product may be plodded, cut, and pressed.

A patent for detergent bars discloses the following:[4]

50–70 parts urea
20–30 ,, amide—from a group consisting of N-methyl-sorbityl myristic amide (Atlas)
30–40 parts sodium lauryl sulfate

Procedure:

Heat amide and urea together until reaction takes place and a clean melt is formed. Mix in the sodium lauryl sulfate to melt. Cool, knead until the mixture is sufficiently stiff to be formed into bars.

Alternative Formula

U.S. Patent 3 057 805 also covers a synthetic detergent cake and process for making it. The patent states that the product contains no undesirable binders, but it includes starch, which is not permitted by government specification P-S-617, and is also omitted from *Tentative Specification* of ASTM Committee D12 (see below).[5]

The Wittwer patent includes alkylaryl synthetic detergent (10–90%), and the stearic acid ester of sodium hydroxyisethionate (90–10%) dry weight, which are synergistic. Any of the conventional additives or perfume may be included, but the combined weight of detergent and binder should be at least 20 per cent.

Procedure:

The ingredients are placed in a mixer, 15–20% water is added and the temperature is maintained at 80–100°C until the mix is a uniform, fused slurry and the water has been reduced to 5 per cent. It is then removed to be formed into bars. Twenty-one compositions are cited, e.g.:

	1	*2*	*3*
Sodium stearoyl isethionate*	65	65	27
Sodium lauroyl ,,	–	20	27
Sodium dodecylbenzenesulfonate	455	65	6
Sodium lauryl sulfate	–	19	–
Corn starch	104	455	–
Tetrasodium pyrophosphate	13	13	–
$Al_2(SO_4)_3$	13	13	–
Total dry weight	650	650	60

* Drucal S; Supplier: 44

It is held by some authorities that those products containing alkylaryl sulfonate are harsh. The use of these sulfonates, which are not biodegradable, will be discontinued because most manufacturers are substituting "soft detergents" which, being entirely biodegradable, avoid pollution of rivers and relieve sewage-disposal plants from this nuisance.

Soap-Surfactant Combination

The following will illustrate the inclusion of several ingredients which were not used in the previous formulas:[6]

No. 1

	A	*B*
Sodium dodecylbenzene sulfonate	41.3%	–
Sodium lauryl sulfate	15.0	–
Coconut acid ester sodium isethionate	–	47.3%
Sodium tallow cocoate (80:20)	30.0	15.0
Stearic acid	–	30.0
Calcium stearate	6.0	–
Titanium dioxide	1.5	1.5
Polyethylene oxide gum	2.0	2.0
Water	4.0	4.0
Perfume	0.2	0.2
	100.0%	100.0%

No. 2 [7]

Sodium higher F.A. monoglyceride sulfate	12%
Sodium dodecylbenzenesulfonate	6
Igepon TE 42 *	5
Sodium soap (3:1 coconut oil:tallow)	59
Total sodium sulfate and sodium chloride	3
Moisture	11
Other ingredients	4
	100%

* Sodium higher F.A. amide of N-taurine.

Alrosol (Geigy), to the extent of 15–25%, may be compounded with toilet soap in the crutcher and then moulded into hard white bar soap of attractive appearance. It will give outstanding detergency in hard and soft water.

Other products of this type contain a high proportion of materials that serve solely to provide desirable physical characteristics to the bar. Such materials include calcium and magnesium stearates, monoglycerides, carbowax 4000, starch, and cellulose derivatives. [8]

Two nationally marketed combination bars are Zest (Procter & Gamble) and Praise (Lever Bros.). Zest contains alcohol sulfates and monoglycerides. Dove (Lever Bros.), which contains some soap, does not qualify as a combination bar. Colgate's Vel Beauty Bar is 100% synthetic. [9,10,11]

To cover such products, the American Society for Testing and Materials (Committee D 12) has offered the following:

Tentative Specifications for
COMBINATION BAR SOAP
ASTM Designation: D2179-63T

Scope:

1. These specifications cover bar soap containing soap and detergent surfactant as the major active ingredients. This bar is intended for toilet use in soft and hard water.

General Requirements:

2. Combination bar soap shall be in bar form, of firm consistency, uniform texture and color, and pleasantly scented. It shall lather freely when used with water having a hardness of 300 parts per million, calculated as $CaCO_3$.

Chemical Composition:

3. Combination bar soap shall conform to the following requirements as to chemical composition. The percentage of each constituent shall be calculated on the basis of the bar as received.

Moisture	14.0%
Detergent surfactant, min	15.0
Anhydrous soap, as soda soap,* min	40.0
Free alkali, as NaOH, max	0.05
*p*H of 1% solution, max 10.5, min 8.5	

* Although calculated as soda soap, other metallic soaps may be present.

Basis of Purchase:

4. The material shall be purchased by net weight.

(a) Deliveries containing more than 14% moisture shall be rejected without further test.

Methods of Sampling and Analysis

5. The material shall be sampled and analyzed in accordance with the Methods of Sampling and Chemical Analysis of Soaps and Soap Products (ASTM Designation: D 460). 1. The purchaser reserves the right to use any additional available information to ascertain whether the material conforms to the specifications.

REFERENCES

1. B. M. Milwidsky, Spray Drying Synthetic Detergents, *Soap Chem. Specialties*, October–December 1962.
2. USP 3 024 258, Continuous Sulfonation: Richard J. Brooks and Burton Brooks to Chemithon Corporation.
3. Ger. Pat. 1 138 497, Skin Protecting Detergent: H. F. Drew and Jack G. Voss to (P&G) Oct. 25, 1962, U.S. Appl. April 20, 1959.
4. F. E. Boettner and J. L. Rainey, USP 2 982 737; to Rohm and Haas.
5. G. G. Wittwer, USP 3 057 305.
6. *Drug Cos. Ind.*, 82 (2), February, 1958.
7. M. Anstetto, USP 3 076 766; to Colgate Palmolive Co.
8. British P. 885 332, 905 952.
9. USP 2 999 068, Personal Use Detergent: to Procter & Gamble Co. Zest, USP 3 024 273; Procter & Gamble.
10. Dove, USP 2 792 349, 2 894 912; Lever Brothers. (*Soap Chem. Specialties*, April 1963.)
11. T. F. Clippinger, USP 3 001 948; to California Research Corp.

chapter 9

ANALYSIS OF OILS AND DETERGENTS

GENERAL

Oils and fats are analyzed by standard methods adopted by the Fat Analysis Committee (F.A.C.) of the American Chemical Society or by the official methods of the American Oil Chemists Society (A.O.C.S.). The tests that follow are abridged from these official methods, with slight modification by the author. For routine commercial analysis, they will be found quite satisfactory.[1]

TEST FOR MOISTURE

Weigh a sample of 5 g into a shallow glass dish, 6 to 7 cm diameter and 4 cm deep, and heat it to constant weight in a vacuum oven repeatedly for 1 hour at 20–25°C above the boiling point of water at 100 mm of mercury.

Alternately, heating in an air oven at 100–105°C or one-half hour, until the loss in each heating period is not greater than 0.05%, is sufficiently accurate for most industrial laboratory work:

$$\frac{\text{Loss in weight} \times 100}{5} = \%\ \text{Moisture}$$

The air-oven method applies to all oils except those of the coconut group and oxidizing oils.

TEST FOR FREE FATTY ACIDS

Dissolve a 5-g sample in hot neutral alcohol (pink to phenolphthalein) or No. 30 Special Denatured Alcohol. Titrate this with standard sodium hydroxide solution.

$$\frac{\text{ml NaOH} \times \text{normality} \times 28.2}{5} = \%\ \text{Free Fatty Acids, as Oleic}$$

An alternate method, which does not require weighing, is to heat the oil or fat to 150°F; then pipette an 8-ml sample into 50 ml of hot neutral ethanol or isopropanol, and titrate with N/4 sodium hydroxide, using a few drops of a 1% alcoholic solution of phenolphthalein as indicator. The test can be performed in a 4-oz oil-sample bottle. When pink end-point is reached, shake the bottle vigorously, until the pink remains 30 seconds.

Reading on burette = % Free Fatty Acids (as Oleic)

Test for Impurities

The presence of impurities in the oil or fat under examination can usually be seen with the naked eye, when the fat is hot, as it is being removed from the oven during the test for moisture. If only a trace of dirt or fat-insoluble matter is present, it may be disregarded. If the amount is considerable or when testing house grease or bone grease that may contain lime soap—soluble in hot fat—the sample used for determining moisture is dissolved in 200 ml of petroleum ether and set away in a cool place for 1 hour. Then the oil is filtered through a Gooch crucible or weighed filter paper.

$$\% \text{ Impurities} = \frac{\text{Increase in weight} \times 100}{5}$$

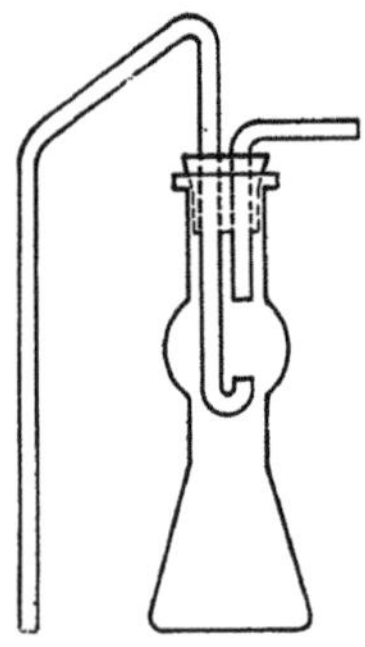

Fig. 9:1. Stokes Flask

Test for Unsaponifiable Matter

Boil a sample of 5 g under reflux condenser with 30 ml of redistilled ethanol and 5 ml of 50% aqueous potassium hydroxide for 1 hour. Using ethanol transfer this to a 4-oz oil bottle or Stokes flask (Fig. 9:1), add an equal volume of water, cool it, and extract it five times with

25 ml of petroleum ether, shaking the solution vigorously after each extraction. Allow the extract to come to the surface and by means of a bent glass tube, held in the cork of the Stokes flask, blow the supernatant extract, as from a wash bottle, into a 500 ml pear-shaped separatory funnel. Repeat this four times.

Wash the petroleum ether extracts twice with 50% ethanol. Filter the extract into a weighed Soxhlet flask, and distill or evaporate it. The residual petroleum ether may be evaporated by setting the flask in a hole of the steam bath for about 15 minutes and then drying in the oven for another 15 minutes. Cool the residue in a desiccator and weigh it.

$$\frac{\text{Gain in weight} \times 100}{5} = \%\ \text{Unsaponifiable}$$

The sum of the moisture, impurities, and unsaponifiable matter is called MIU.

Test for Titer

The titer test is the determination of the highest point in °C reached during the solidification of fatty acids of an oil or fat, under conditions explained as follows: The test will give the soapmaker an idea of the character of the soap he will obtain.

Dissolve 250 g of C.P. caustic potash in 1000 ml of C.P., or dynamite grade, glycerol. To avoid foaming, do not heat over 140°C.

Heat 80 ml of the glycerol/caustic solution to 150°C, and add 50 ml of melted fat. Stir the mixture well and continue heating at 140–150°C until the soap becomes fairly viscous, clears, and thins out again. This should require at least 15–20 minutes. Pour a bit of the soap into a test tube, add hot distilled water and stir until solution takes place. The solution should be very clear. If so, it may reasonably be assumed that the saponification is complete.

Allow the mass to cool somewhat, add hot water to dissolve the soap, heating to boil if necessary. Add cautiously, 50 ml of 30%-sulfuric acid, and boil until the clear fatty acids separate out. The aqueous solution must be acidic to break the soap completely. If insufficient acid is used, the entire soap may not be decomposed and the fatty acids will not be clear. (The whole operation may be carried out in an 8-inch evaporating dish.) Cool somewhat, syphon off the acid water, and wash the fatty acids several times with hot water.

Transfer the fatty acids to a 100-ml beaker, taking care that in decanting, no water runs into the beaker. Heat to 130°C as rapidly as possible, while stirring with the thermometer. Allow solution to cool to 2–3 degrees above the expected titer.

Transfer to a titer test tube (100 mm × 25 mm), which is set into a perforated cork of a 16-oz salt-mouth bottle, so that the test tube will be held rigidly. Suspend the standard titer-test thermometer (graduated in 0.2°C), so that the bulb reaches the center of the fat in the test tube. Stir with the thermometer rapidly, while the mercury drops, then remains stationary, and begins to rise gradually. Stop the stirring and observe the highest point to which the mercury rises.*

This figure is recorded as the *titer*.

Tests for Color

There are several systems by which color of oils and fats is measured. For vegetable oils, fairly light in color, the American Oil Chemists' Society's official method CC13 b-45 has been adopted. The color is determined by matching a column of oil, 133.35 mm (5.25 in.) high, in a tintometer, using red and yellow standard Lovibond color glasses for comparison.

The standards for red glass are determined by the National Bureau of Standards, in terms of the Priest-Gibson N · color scale. The yellow glasses are not standardized because there is little visual difference between the glasses produced by the different manufacturers. The Wesson tintometer is equipped with 31 Lovibond glasses.

For refined oil: Use only 1 yellow glass, 35 yellow for refined cottonseed oil and refined peanut oil; 70 yellow for refined soybean oil. Use not more than two red glasses up to and including 13 red, and not more than 3 red glasses above 13 red.

For refined bleached oil, the ratio of yellow to red is:

Oil	*Yellow*		*Red*
Cottonseed, peanut, corn	10	to	1–3.5
	35	for	3.5 and up
Coconut, palm kernel	6	to	1–3.9
	10	to	1 for 3.9 and up
Soybean	10	to	1–3.5
	70	to	3.5 or above
Tallows, greases, fatty acids:	10	to	1–3.5
	35	for	3.5–5 inclusive
	70	for	Above 5

*This test has been automated so that three samples may be run simultaneously. An electric apparatus stirs three thermometers at the same time; (Fisher Titer Tester).

The color of dark oils may be read by using a column of oil, 25.4 mm (1.25 in.) high, and matching with yellow and red glasses, making note that the reading was taken on a column of 25.4 mm. Specifications covering grades of crude and refined oils are published by the National Cottonseed Products Association.

The test for color used by soapmakers, and also the official method of the American Oil Chemists' Society, applicable to animal fats and all fats and oils too dark to be read by the Wesson Tintometer, is as follows:

Fat Analysis Committee Color Test

In the *A.O.C.S. Methods of Analysis* (Revised, 1964), the fat to be tested is heated to about 10°C above the melting point. It is then poured into a small test tube and matched with the LaPine Color Standards. The latter consists of a 3-Aperture Color Grading Comparator, with four discs of the F.A.C. color scale. Discs are also supplied for Gardner Color Standards, for evaluating color of varnishes, oils, lacquers and resins. The paint industry has adopted the Gardner standards.*

Tubes of the same size as the standards, filled with the oil to be graded, are matched to the nearest color standard.

Other Systems

The National Petroleum Association uses a color system which corresponds numerically to the system of the ASTM and Union colorimeter.

The Hardy recording spectrophotometer** permits exact measurement of color and thus eliminates possibility of visual error. However, the instrument is expensive and colors cannot be expressed in single figures as yellow and red.

The various systems for grading color have been correlated in a chart (Fig. 9:2), showing comparative readings for the different scales. It may be obtained free of charge.

Iodine Number (Wijs Method)

The iodine number or iodine value of an oil is expressed as the number of centigrams of iodine absorbed by 1-g sample (% iodine absorbed). It is also a measure of the unsaturation of an oil or fat.

* Produced by H. A. Gardner Laboratory, Washington, D. C.

** General Electric Company.

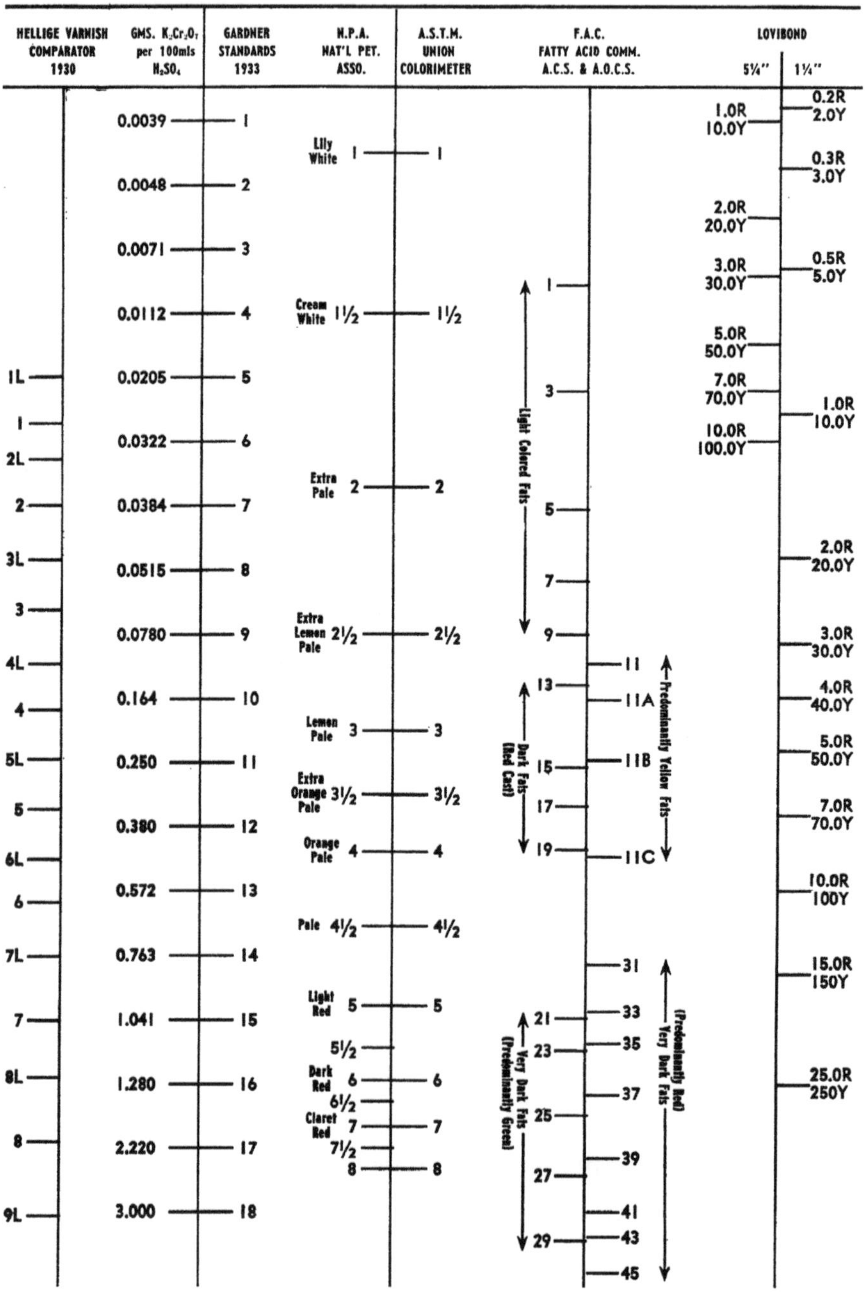

FIG. 9:2. COMPARATOR FOR COLOR STANDARDS
(*Harchem Div., Wallace & Tiernan, Inc., Belleville, N,J,*)

The higher the iodine number, the more unsaturated is the oil, i.e., the more readily will it oxidize and vice versa.

Preparation of Reagents

Wijs Iodine Solution:

The Wijs method calls for the solution of 13 g resublimed iodine in 1 liter of glacial acetic acid. Dried chlorine gas is added to the solution until the original titration is not quite doubled.*

N/10 Sodium Thiosulfate:

Dissolve 24.8 g sodium thiosulfate ($Na_2S_2O_3 \cdot 5H_2O$) and dilute it to 1 liter at the temperature at which the titrations are to be made.

Starch Paste:

Boil 1 g of starch in 200 ml of distilled water for 10 minutes and cool to room temperature.

Potassium Iodide Solution:

Dissolve 150 g of potassium iodide in distilled water and make up to 1 liter.

Potassium Dichromate Solution:

Dissolve 0.9658 g of potassium dichromate in distilled water and make up to 250 ml. This solution is equivalent to 0.01 g iodine/ml. In laboratories where the iodine number is determined frequently, one may make up a liter at a time, using four times the quantity of potassium dichromate.

All the reagents should be chemically pure and of a standard recommended by the American Chemical Society (A.C.S. Grade).

To Standardize Sodium Thiosulfate Solution

With a pipette, place 25 ml of potassium dichromate solution in an Erlenmeyer flask. Add 10 ml of potassium iodide solution and 5 ml of hydrochloric acid. Dilute with 100 ml of distilled water and titrate with the sodium thiosulfate until the solution turns straw yellow. Add a few drops of starch paste and continue the titration until the blue color disappears. Note the number of ml required to discharge the color.

---

* The preparation of this solution is a very disagreeable job and, therefore, the author suggests that the standard solution be purchased from a supply house.

To Determine Iodine Value

Weigh into a dry 250-ml iodine flask (or an 8-oz glass stoppered bottle) containing 20 ml of re-distilled carbon tetrachloride, 0.1 to 0.5 g of oil or melted fat, using the lower quantity for oils with high iodine values. Add 25 ml of Wijs solution by means of a pipette, allowing it to drain for a definite time. The excess iodine should be 50–60% of the amount added; that is, 100–150% of the amount absorbed. Moisten the stopper with 15%-potassium iodide solution to prevent loss of iodine or chlorine, but guard against an amount sufficient to run down inside the flask. Stopper the flask and set it in a dark place. Prepare two blanks identical with the sample, but without the fat or oil. Let the flasks stand in the dark for 30 minutes at room temperature.

Remove the flasks from storage and add 20 ml of potassium iodide solution and 100 ml of distilled water. Titrate with sodium thiosulfate solution—which is added gradually with constant rotary motion of the flask—until the yellow color has almost disappeared. Toward the end of the titration, stopper the flask and shake it vigorously, so that any iodine that may still be dissolved in the carbon tetrachloride may be taken up by the potassium iodide solution. Finish this titration, adding a few drops of starch paste. Discharge any blue color still remaining. Then titrate the blanks.

Variations in temperature appreciably affect the titer of the iodine solution, as acetic acid has a high coefficient of expansion. It is therefore essential that the blanks and the determination on the sample be run at the same time. The number of ml of standard thiosulfate solution required by the blanks, less the amount used in the determination, gives the thiosulfate equivalent of the iodine absorbed by the sample. Calculate to centigrams of iodine absorbed by 1 g sample (% iodine absorbed).

Example: Sample, 0.2264 g of olive oil

	$Na_2S_2O_3$ ml
The blank requires	27.65
Back titration on sample	17.45
The difference is	10.20

Therefore, $\frac{10.2}{0.2264}$ = ml of thiosulfate for a gram of olive oil.

The 25 ml of potassium dichromate, equivalent to 0.25 g of iodine, requires 13.8 ml of thiosulfate for titration. Consequently, 1 g of iodine will require four times 13.8 ml, or 55.2 ml of thiosulfate. Therefore—

$$\frac{10.2}{0.2264} \Big/ 55.2 \times 100 = 81.6\% \text{ of iodine absorbed,}$$

which is designated *iodine value* or *iodine number.*

Saponification Number

The saponification number or Koettsdorfer Number, of an oil or fat is the number of milligrams of potassium hydroxide required to saponify 1 g of the sample.

The saponification equivalent corresponds approximately to the molecular weight, in case of fatty acids or esters of monohydric alcohols. For glycerides, the saponification equivalent multiplied by 3 corresponds to the molecular weight.

$$\text{Saponification Equivalent} = \frac{56{,}100}{\text{Saponification Number}}$$

Preparation of Reagents

N/2 Hydrochloric acid:

Carefully standardized.

Alcoholic Potash:

Prepare the alcohol (S.D. 30 or 34) by boiling 1500 ml with 10 g of potassium hydroxide, under a reflux condenser, on a steam bath, for one hour and distil off 1 liter. Dissolve 27 g of potassium hydroxide in 1 liter of distilled alcohol, without heating. The solution should be clear.

Alcoholic phenolphthalein:

Prepare a 1% solution of phenolphthalein in alcohol.

To Determine Saponification Number

Weigh accurately about 5 g of the filtered sample into a 150-ml Erlenmeyer flask. Pipette 50 ml of alcoholic potash into the flask, allowing it to drain a definite time. Connect the flask with an air condenser and boil gently for 30 minutes. Cool and titrate with N/2 hydrochloric acid, using phenolphthalein as indicator. Run a blank simultaneously.

Calculate saponification number as follows:

$$\frac{\text{(Titration of blank-Titration of sample) } 28.05}{5} = \text{Saponification Number}$$

This holds when exactly 5 g are taken for the determination otherwise, use the weight of the sample as the divisor.

Melting Point of Fats

There are several methods for determining the melting point of natural fats. Where two chemists are checking each other, there should be a mutual agreement previously, on the method to be employed. Fats and oils are not pure organic chemicals and therefore, their melting points are not well defined. In melting, they go through an opalescent stage and finally turn clear.

Capillary Method

The fat to be tested should be clean and absolutely dry. The hot fat may be run through a filter paper. The capillary tubes, 1 mm in diameter and 80 mm long, are immersed in the hot fat to a distance of 10 mm. The tubes are then sealed at the bottom, by holding them in a bunsen flame, taking care not to burn the fat. Three capillary tubes are set in the refrigerator at 40°F overnight. The tubes are then attached to the titer thermometer, by means of rubber bands, so that the bottom of each tube is flush with that of the thermometer bulb.

The bulb and tubes are suspended in a 600-ml beaker of distilled water to a depth of 30 mm. The water is heated with a bunsen flame so that the temperature will rise 1° a minute. When the fat turns clear, the thermometer is read and the average of the three melting points taken. This should be within 0.5°.

Acid Number

The acid number of an oil or fat is expressed as milligrams of potassium hydroxide required to neutralize 1 g of the sample.

Weigh 5–20 g of the sample into a 4-oz oil bottle, add 50 ml of hot ethanol, previously neutralized with N/10 sodium- or potassium-hydroxide, using phenolphthalein as indicator. Titrate until the pink color persists for half minute. Calculate milligrams of potassium hydroxide required.

$$\frac{\text{ml} \times \text{Normality} \times 0.056 \times 1000}{\text{Weight of sample}} = \text{Acid Number}$$

Ester Value = Saponification Number minus Acid Number

The ester value indicates the approximate amount of glycerides present.

Acetyl Value

The acetyl value is the number of milligrams of potassium hydroxide required to neutralize the acetic acid obtained on saponification of 1 g of acetylated fat.

The determination is based on the principle that glycerides that contain hydroxylated fatty acids assimilate, on heating with acetic anhydride, one or more acetyl groups, according to whether the fatty acids contain one or more hydroxyl groups. In this reaction, the hydrogen atom of the alcoholic hydroxyl group or groups is replaced by the acetic acid radical.

To Determine Acetyl Number

Boil 10–50 g of a sample with an equal volume of acetic anhydride, in an acetylation flask, for two hours. Then pour the mixture into a large beaker containing 500 ml of water and boil this for 30 minutes. To prevent bumping, pass a slow current of carbon dioxide through a finely drawn-out tube that reaches the bottom.

Allow the mixture to separate into two layers. Syphon off the water, and boil the oily layer with fresh water until it is no longer acid to litmus. Separate the acetylated fat from the water; dry and filter the fat while it is kept hot in the drying oven.

Saponification

Follow the method prescribed under *Saponification Number*, and calculate in the usual way the milligrams of potassium hydroxide required to saponify 1 g of sample.

The acetyl value is calculated from the following formula:

$$A = \frac{S' - S}{1 - 0.00075S}$$

where A is acetyl value
S′ is saponification after acetylation
S is saponification number before acetylation.

Cold Test

Place 28 ml of the oil in a cold-test bottle or 4-oz oil bottle. Set the bottle in a freezing mixture consisting of shaved ice and salt. Immerse a cold-test thermometer in the oil and hold it in the bottle by means of a cork gasket. When the oil becomes solid, let it stand for half an hour longer. Then remove the bottle from the freezing mixture and stir it with the thermometer as soon as the oil has softened sufficiently. Grasp the top of the bottle so that the heat of the hand does not warm the oil and tip it at an angle of 45° below the horizontal.

Note the temperature at which the oil will just run from one end of the bottle to the other. This point, expressed in Fahrenheit degrees, is taken as the cold test.

The cold test of an oil is a measure of how well it will lubricate at low temperatures. Therefore, the lower the cold test, the better the lubricity at low temperature.

Emulsions utilizing neatsfoot oil of low cold-test are more fluid and strike through leather more easily, e.g., in fat-liquoring.

Smoke, Flash, Fire Tests

The smoke, flash, and fire points of animal and vegetable fats and oils are determined by heating the dry oil or fat under standard conditions in the Cleveland open cup. They are expressed in Fahrenheit degrees.

Smoke Test

Fill the cup to the mark with the sample to be tested, and then heat it slowly over a bunsen flame to within 75°F of the expected smoke point.

The cup, with the thermometer hanging in the center of the melted fat or oil, should be suitably shielded by means of a box, constructed of asbestos board, which supports an electric bulb on one side, at the level of the surface of the oil. When a bluish smoke appears, the temperature is noted. This is the *smoke point.*

Flash Test

Fill the cup with the molten fat to be tested, and suspend a thermometer so that the bulb is about 6 mm from the bottom of the cup. Heat slowly, to raise the temperature to within 100°F of the expected flash point, and then regulate the heat so that the rise will be about 10°F a minute. By means of a blowpipe, light a gas flame so that it will be about ⅛-inch in diameter. As the oil is heated draw the flame of the blow-pipe across the surface of the oil at intervals of one-half minute.

At first, a bluish halo may appear as the flash point is approached. The real flash point is recorded when the flash gives a yellow flicker and goes out again.

To Determine Fire Point

After noting the flash point, continue the heating until a point is reached, when touching the surface of the oil with the test flame causes it to burn for at least five seconds.

Thiocyanogen Value

The thiocyanogen value $(SCN)_2$ is a measure of the unsaturation of fats and oils (see Chapter 4), expressed in terms of the equivalent centigrams (%) of iodine absorbed by 1 g of sample.

Thiocyanogen reacts to the same extent as iodine with oleic acid, but adds to only one bond of linoleic acid and to two bonds of linolenic acid. The method permits the determination of oleic, linoleic, and linolenic acids, provided that other unsaturated acids are not present, and that the iodine number and amount of saturated acids are known.

The method requires 24 hours to complete and is quite complicated, requiring considerable analytical skill.

Volatile Fatty Acids

The volatile fatty acids are those that can be separated from the hydrolyzate of a fat or oil by steam distillation. Butter fat, the lauric acid oils, and a few others, containing up to twelve carbon atoms, yield volatile fatty acids.

Although no method has as yet been devised to separate all the volatile from the non-volatile fatty acids, the Reichert-Meissl method gives an approximate determination of the water-soluble volatile fatty acids, and the Polenske method may be used to determine the insoluble volatile fatty acids, with reproducible results, if the methods are followed carefully.

The *Reichert-Meissl value* is defined as the number of ml of N/10 potassium hydroxide solution required to neutralize the soluble fatty acids from 5 g of sample.

The *Polenske number* is defined as the number of ml of N/10 potassium hydroxide solution required to neutralize the insoluble fatty acids from a 5 g sample.[2]

To Determine Reichert-Meissl Value

Weigh a 5 g sample into a clean dry 300 ml flask (Fig. 9:3). Add 20 ml glycerol sodium hydroxide solution (made by adding 20 ml of 50%-sodium hydroxide to 180 ml of pure glycerol). Heat over asbestos-covered wire gauze until completely saponified, shown when the mixture becomes clear. Shake the mixture by rotating the flask to keep the foam down. Add 130 ml of recently boiled cold water and add slowly 10 ml of dilute sulfuric acid (200 ml of concentrated acid diluted to 1000 ml with water). Add 1 g of powdered pumice, which has previously been strongly ignited and cooled under water until used. Distill without previously melting the separated fatty acids.

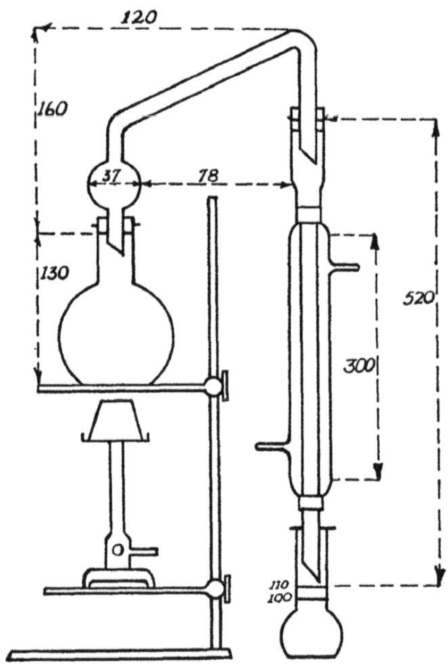

FIG. 9:3. APPARATUS FOR DETERMINING REICHERT-MEISSL AND POLENSKE VALUES

All the apparatus should be set up as in Figure 9:3. The 300-ml distilling flask should rest on a piece of heavy asbestos board, provided with a 50-mm hole in the center. The flame should be so regulated that 110 ml of distillate will collect in as nearly 30 minutes as possible; and the distillate should come over at a temperature not higher than 18–20°C.

When 110 ml of the distillate has been collected, remove the flame and substitute for the receiving flask a 25-ml cylinder to catch any drops of distillate that may fall later. Mix the contents of the flask containing the distillate and immerse the flask almost completely in water at 15°C for 15 minutes.

Filter the distillate through a dry 9-cm filter paper. Titrate 100 ml of the filtrate with N/10 potassium hydroxide solution, using 6 drops of 1% alcoholic solution of phenolphthalein. The pink color should persist for 2 minutes. Make a blank determination, using the same proportion of reagents as in the actual determination.

1.1 (Titration of sample minus Titration of blank)
= Reichert-Meissl Value

To Determine Polenske Number

Remove the remainder of the soluble acids from the insoluble acids on the filter, by washing with three successive 15-ml portions of water previously passed through the condenser, the 110-ml receiving flask, and the 25-ml cylinder. Dissolve the water-insoluble acids with three successive 15-ml portions of neutral 95% ethanol through the filter, each portion having previously been passed through the condenser, receiver, and 25-ml cylinder.

Titrate the combined alcoholic washings with N/10 potassium hydroxide, and calculate the Polenske number according to the definition given previously. All determinations should be made in duplicate. All details must be followed precisely to obtain satisfactory results.[2]

SOME SPECIAL TESTS

Bleach Test on Animal Fats

Weigh 50 g of the sample into a 100-ml tall-form beaker, and heat it to 105°C until most of the moisture is driven off. Add 2 g of A.O.A.C. standard fuller's earth. Agitate with the thermometer for 15 minutes, keeping the temperature of the fat between 100 and 105°C. Filter through dry filter paper, using a hot funnel if necessary, until a sufficient quantity has been filtered, for reading color.

Specific Gravity of Oils

The specific gravity of an oil should be determined at 25°C with a pycnometer. For close approximation, the hydrometer may be used.

The so-called weight-per-gallon pycnometer is an instrument constructed of metal, whose volume is 83.3 ml. When filled with a liquid, the net weight in grams, divided by 10, is the weight in pounds per gallon. We have found this instrument very useful as a laboratory tool.

The Westphal balance is also a very useful instrument in a soap and chemical specialties laboratory. The appropriate weights or riders on the beam may be used for determining the specific gravities of liquids lighter or heavier than water.

Test for Marine Animal Oils in Vegetable Oil

This test is applicable only in the absence of metallic salts.

Dissolve about 6 g of the sample in a test tube, in 12 ml of a mixture of equal parts of chloroform and glacial acetic acid. Maintain the solution at 20°C and add bromine, drop by drop, until a slight excess is indicated by brown color. Let it stand 15 minutes and then place the test tube in boiling water. If only vegetable oils are present, the solution will become perfectly clear. In the presence of fish oil, the solution will be cloudy or show a precipitate, due to the presence of insoluble bromides.*

There are several qualitative tests to determine the presence of certain oils in a mixture. For example, olive oil, which is an imported product and is generally higher in price than domestic oils, has at times been found to be adulterated with less expensive oil.

Halphen Test for Cottonseed Oil

Although the Halphen test is supposed to detect the presence of cottonseed oil in admixture with other oils and fats, there are many exceptions in which the test becomes inoperative.

Kapok oil reacts with equal if not greater intensity than cottonseed oil and fats from animals which have been fed with cottonseed meal often respond to this test.

Hydrogenated cottonseed oil, or even oil heated to 250°C, interferes with the reaction so as to make it indefinite.

For carrying out this test, mix carbon disulfide (containing 1% of sulfur in solution) with an equal volume of amyl alcohol. In a large test tube, mix 10 ml of the sample with an equal volume of the reagent. Set the test tube in a steam bath to boil off the carbon disulfide and then immerse it in a bath of boiling brine for one to two hours. A red color indicates the presence of cottonseed oil. The greater the quantity of cottonseed oil present, the sooner will the red color appear.

* The author's observation has been that this test does not show the presence of hydrogenated whale oil in tallow.

Baudoin's Test for Sesame Oil

To 10 ml of oil in a test tube, add 2 drops of a 2%-alcoholic solution of furfural and 10 ml of concentrated hydrochloric acid. Shake well. In the presence of sesame oil, the aqueous liquid will become crimson in color. This test is sensitive to 1% of sesame oil.

The color reaction is due to the presence of sesamin and sesamoline in the unsaponifiable portion of the oil.

Bellier's Test for Peanut Oil

Preparation of Reagents

Alcoholic potash:

Dissolve 8.5 g of potassium hydroxide in 70%-alcohol, and make up volume to 100 ml.

Acetic Acid:

Dilute 120 ml of 36% acetic acid to 150 ml. 1.5 ml of this solution should completely neutralize 5 ml of the alcoholic potash.

Procedure

Weigh 1 g of sample into a dry test tube, add 5 ml of alcoholic potash; and boil gently over a small flame, holding the tube with a holder until the oil is completely saponified, avoiding evaporation as far as possible. This requires about 2 minutes.

Add 1 ml of acetic acid, mix well, and cool for 30 minutes in tap water, shaking the tube occasionally. Then add 50 ml of 70%-alcohol, containing 1% by volume of hydrochloric acid (sp. gr. 1.16). Shake well and again place the tube in water for an hour.

If no peanut oil is present, a clear opalescent liquid is formed; if more than 10% is present, a flocculent crystalline precipitate remains. Even with 5%, a distinct precipitate remains and separates out on standing.

Fitelson Test for Teaseed Oil
(Lieberman-Burchard Modification)

Of all the oils that resemble olive oil, none matches it so well in chemical and physical constants, as teaseed oil. It is for this reason, that olive oil has been occasionally adulterated with it, or it has been substituted for olive oil.

Qualitative Test

Measure into a test tube (approximately 18 × 150 mm) exactly 0.8 ml of acetic anhydride, 1.5 ml of chloroform, and 0.2 ml of concentrated sulfuric acid. Mix and cool in water to 5°C. Add 7 drops of oil. (For this, ordinary glass tubing, 4 mm outside diameter and 2 mm inside diameter, is used.) These 7 drops will weigh approximately 0.22 gram.

If the solution of the oil and reagent is cloudy, add acetic anhydride, drop by drop, shaking after each addition until the turbidity disappears. Maintain at 5°C for five minutes. Add 10 ml of cold ether, cork the tube, and invert it. Place it in ice-water and observe the color.

Pure teaseed oil gives an intense red color within a minute and then fades. The depth of color is proportional to the amount of teaseed oil present. By making up known mixtures of olive oil and teaseed oil and subjecting them to the same reaction if the sample contains at least 10% of teaseed oil, a fair estimate of the amount of teaseed oil present in a suspected sample can be made.

Dark olive oil or industrial oils sometimes mask the color. In the latter case, 5 g of the sample, mixed with 5 g white mineral oil, is saponified by boiling for 10 minutes with 5 ml of 50%-potassium hydroxide in about 30 ml of ethanol. This is poured into a separatory funnel. An equal volume of water is added and the mixture is shaken. When the layers separate, the aqueous portion is drawn off and the oily portion is washed several times with water. The oil is then dried with anhydrous sodium sulfate, filtered through dry filter paper, and the test for teaseed oil is made, as indicated before.

KREIS TEST FOR RANCIDITY

Pour 5 ml of the oil and 5 ml hydrochloric acid (free from nitrosyl chloride) in a test tube. Stopper and shake well for 30 seconds. Add 5 ml of 0.1% solution of phloroglucinol in ether. Shake mixture again for 30 seconds and allow it to stand for 10 minutes. A good pink color, appearing in the lower acid layer, indicates rancidity. Pale orange, yellow, or slight pink should be disregarded. If the acid layer becomes pink, make a mixture of 1 part of the sample with 9 parts of liquid petrolatum and another mixture of 1 part of the sample and 19 parts of liquid petrolatum. Test both mixtures as before (see page 118).

Test for Hardness in Water

Hardness of water is determined by titration of a measured amount of water, with a standard soap solution. The point at which a good lather is maintained for five minutes is noted on the burette and the hardness is calculated from that.

The actual procedure is as follows:

The reagent used, is the Standard Soap Solution, prescribed by the American Public Health Association, one ml of soap solution is equivalent to 1 mg of $CaCO_3$.

This value is based upon the test of a 50-ml sample of standard $CaCl_2$ solution titrated at 72°F to a permanent foam end point. *Permanent Foam* is defined as a lather persisting for five minutes, with unbroken surface the, bottle lying on its side.

Test: Transfer 50-ml sample to an 8-oz round bottle. Add standard soap solution from a burette (0.2–0.3 ml), at a time and shake vigorously after each addition. Lay bottle on its side and observe the time required for a break to form on the surface. Continue additions until a lather persists, unbroken for five minutes.* Multiply the result by 20. That represents parts per million of $CaCO_3$ or hardness. To read grains per gallon, use 58.3 ml sample. Deduct lather factor from burette reading. The corrected burette reading is directly equal to grains $CaCO_3$ per gallon. That is, each ml soap solution equals 1 gr/gal hardness.

For other tests on water, consult *Standard Methods of Water Analysis*, American Public Health Association, 1790 Broadway, New York, N.Y., 10019.

ANALYSIS OF SOAP AND SOAP PRODUCTS

Sampling and Preparation of Sample

This method of sampling is applicable to bar soap, soap powders, cleansers, flakes, paste, and liquid soaps.[3]

Product is sampled as convenient, taking a sample of 1% of the packages from lots of one thousand; or, of bar soap, one bar from each ten cases. The samples are placed in airtight containers. Liquids are sampled with a glass tube one-half inch in diameter, which is long enough to reach the bottom of the drum or barrel. Pastes are sampled with a *trier*, which is a half-round metal tube, long enough to reach

* When using prepared Standard Soap Solution bought from a supply house, there is a stated lather factor on the bottle. This is deducted from the burette reading.

the bottom of the package. For bar soap, a portion of each bar is ground through a meat chopper and a composite sample is made up in two Mason jars. One is for analysis and the other for future reference; or for shipment to a referee chemist.

Test for Moisture

The following method is applicable to soap and soap products, except those that contain water-miscible, volatile substances. The

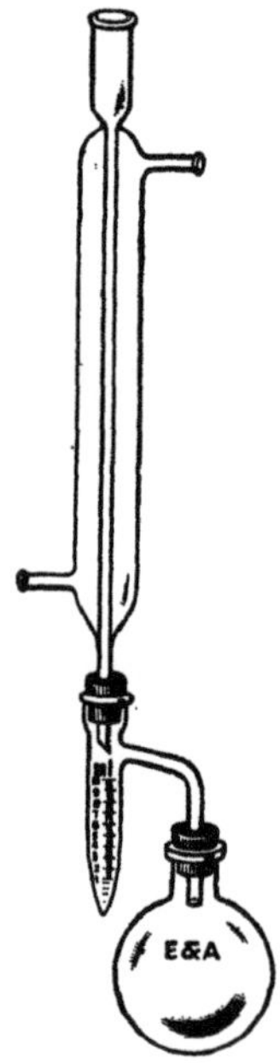

Fig. 9:4. Stark and Dean Distillation Unit

soap is distilled from a flask with xylene or toluene, using a Stark and Dean distillation trap and a reflux condenser (Fig. 9:4).

For soaps containing 5–25% of moisture, use a 20-g sample. For products containing over 25% moisture, use a 10 g sample.

Weigh the sample into a flask, add 50 ml of toluene and 10 g of anhydrous sodium acetate, or 20 g of rosin, to prevent frothing. Also, add a few glass beads or small pebbles.

Distill until no more water drops into the distillation trap. Then allow it to boil for ten minutes more. A spiral copper wire is used to push down the last drops of water that may remain in the condenser, and it is washed down with a few ml of toluene, when cool.

$$\frac{\text{Volume of water} \times 100}{\text{Weight of Sample}} = \text{Moisture } \%$$

Moisture by Air-oven Method

When soaps and other products do not contain large amounts of silicate or over 1% glycerol, or for soaps not produced from linseed oil, which do not oxidize readily, the following method may be used. Otherwise the distillation method is employed.

Weigh accurately, a sample of about 3 grams in a crystallizing dish and place it in the oven at 105°C for four hours, until the weight is practically constant. Cool in desiccator and reweigh.

$$\frac{\text{Loss in weight} \times 100}{\text{Weight of Sample}} = \text{Moisture and Volatile } \%$$

Test for Free Caustic Alkali

Weigh a 5-g sample into a 300-ml tall-form beaker. Dissolve it in 150 ml of neutral ethanol, using phenolphthalein as indicator. Filter with suction and titrate the filtrate with N/10 sulfuric acid.

$$\text{ml N/10 } H_2SO_4 \times 0.004 \times 20 = \%\ NaOH$$
$$\text{ml N/10 } H_2SO_4 \times 0.00561 \times 20 = \%\ KOH$$

Test for Carbonate

The residue on the filter, from the previous determination, is dissolved with hot water and when cool, titrated with N/10 sulfuric acid, using methyl orange as indicator.

$$\text{ml N/10 } H_2SO_4 \times 0.0053 \times 20 = \%\ Na_2CO_3$$
$$\text{ml N/10 } H_2SO_4 \times 0.0069 \times 20 = \%\ K_2CO_3$$

If the soap contains no fillers or builders, sodium and potassium carbonate are the only free alkalies that will ordinarily be found. If fillers and builders are suspected, it is always best to make a qualitative test for sodium silicate and other alcohol-insoluble matter. When the free caustic is being determined and insoluble matter is detected in the alcoholic solution of the soap, this insoluble residue must be analyzed.

Tests for Sodium Silicate

Qualitative Test

Dissolve about 1 g of the sample in 25 ml of hot distilled water. Heavily silicated soaps may show turbidity because some of the silicate has been dehydrated to such an extent that it no longer dissolves. Especially is this true of soap chips and some of the spray-dried powders. There may also be present insoluble fillers such as silex, starch, etc. In the absence of starch or dextrine, take 10 ml of the clear solution and pour it into a test tube containing 10 ml of acetone and 1 ml of a solution prepared by dissolving 10 g of pure sodium aluminate and 2 g of sodium chloride in 1 liter of distilled water. A flocculent precipitate indicates sodium silicate.

In the case of small amounts of silicate, such as may be present in white floating soap or toilet soap, the test tube containing the sample and reagent may be heated in a steam bath for a minute. If silicate is present, a light precipitate appears. This is a very ingenious test which was published in the *Journal of Industrial and Engineering Chemistry*, in 1915.

Quantitative Determination

Ash 1 g of soap in a platinum crucible. Dissolve the ash in hot distilled water and transfer the solution to a 300-ml beaker. Acidify with concentrated hydrochloric acid and a few drops of concentrated nitric acid. Evaporate to dryness and heat on a hot plate around 130°C for an hour. Pick up the silica residue with hot water, again acidify as before, and wash down anything that may have spattered on the sides of the beaker, using a policeman.

Evaporate and heat as before. Add a little water and a few drops of hydrochloric acid and filter through ashless filter paper. Incinerate in a platinum crucible. Heat over a bunsen flame until all traces of carbon have disappeared and the silicon dioxide is perfectly white. Cool in a desiccator and weigh.

$$\text{g } SiO_2 \times 1.2571 \times 100 = \% \text{ sodium silicate } (Na_2Si_4O_9)$$

Tests for Borax

Qualitative Test

To a 1-g sample in a test tube, add 10 ml of dilute hydrochloric acid. Boil until the fatty acids separate out. Cool and filter through paper that has previously been wetted with water.

Immerse a strip of turmeric paper in the aqueous filtrate and allow the paper to dry. It will be colored a deep red if borax is present. If a few drops of ammonia are placed on the test paper, a blue color confirms the presence of borax. The test is sensitive to 0.05% of borax.

Quantitative Determination

To 10 g of soap (5 g if soap contains more than 5% of borax) in a platinum evaporating dish, add 2.15 g of a fusion mixture consisting of 200 g of sodium carbonate and 15 g of silicon dioxide, finely powdered. Add 15 ml of 95% alcohol and mix with a glass stirring rod. After washing off the rod with a few ml of alcohol, evaporate the mass to dryness on a water bath. Ignite gently until all the combustible matter is destroyed, cover the dish with platinum foil, and fuse.

Completely disintegrate the fused mass by boiling with water and transfer the solution to a 250-ml round-bottom flask. Acidify with 20 ml dilute hydrochloric acid (1:1). Heat nearly to boiling and add a moderate excess of dry precipitated chalk. Connect with a reflux condenser and boil vigorously. Filter the precipitate through a folded filter and wash with hot water keeping the total volume of liquid below 100 ml. Return the solution to the flask, add a little chalk and again boil under the reflux. Remove the flame and connect the top of the condenser with a suction pump until the boiling has ceased.

Add 50 ml of glycerol and titrate with N/10 sodium hydroxide, using phenolphthalein as indicator. When the end point is reached, add 10 ml of neutral glycerol and titrate once more. Repeat this until the addition of glycerol does not affect the end point.

$$\frac{\text{ml N/10} \times 0.00955 \times 100}{\text{Weight of Sample}} = \%\ \text{Borax}\ (Na_2B_4O_7 \cdot 10\ H_2O)$$

This method is described in *J. Ind. Eng. Chem.*, 5:645 (1913).

Test for Total Alkali and Total Fat

Dissolve 5 g of soap in hot water. Add 25 ml of N/H_2SO_4 and set the solution on the steam bath until the clear fatty acids float on the surface of the water. Cool and transfer to a pear-shaped separatory funnel with U.S.P. ether. Allow liquids to settle and run the aqueous portion into another separatory funnel. Wash the ether in the first funnel once with water and when settled, combine the washings and shake out with ether. Then filter the ether extracts from both funnels into a Soxhlet flask and distill off the ether. Dry the fatty acids on the steam bath, cool in a desiccator, and weigh.

Fatty extract × 20 = % Total fatty acids and unsaponifiable matter.

The combined aqueous solutions are titrated with N/NaOH using methyl orange as indicator.

$$25 \text{ ml} - \text{NaOH titration} \times 0.031 \times 20 = \%\ \text{Total } Na_2O$$

The total Na_2O includes the alkalinity due to free caustic, carbonate silicate, if any, and all alkaline builders, which must be subtracted to find the percentage of sodium oxide combined as soap.

Conversion Factors:

Multiply	by	to find
Sodium silicate	× 0.20448	= Na_2O
Sodium carbonate	× 0.5849	= Na_2O
Sodium hydroxide	× 0.775	= Na_2O

Test for Sodium Chloride

To the aqueous solution of the preceding determination of total alkali, add a few drops of 10% potassium chromate and titrate with N/10 silver nitrate.

$$\text{ml } AgNO_3 \times 0.00585 \times 20 = \%\ \text{NaCl}$$

Test for Unsaponifiable and/or Unsaponified Fat

Dissolve 5 g of soap in 50 ml of hot water, cool and add 50 ml of ethanol. Transfer the solution to a separatory funnel and shake it with 50 ml of petroleum ether. Allow liquids to settle and draw off the alcoholic layer into a second separatory funnel. Wash the extract with 25 ml of 50% ethanol. Draw off the washings into the second separatory funnel and shake with 25 ml of petroleum ether. Draw off the alcohol and wash the extract as before, and filter both extracts into a 150-ml Soxhlet flask. Distill off the petroleum ether. Dry the residue on the steam bath, cool it in a desiccator, and weigh it.

$$\text{Wt Extract} \times 20 = \%\ \text{Unsaponifiable, Unsaponified fat.}$$

Tests for Stability of Soap

There are several methods for testing the effectiveness of an antioxidant in soap. In the Swift test for stability, the soap is aerated in an oven at 97.7°C. Other methods employ higher or lower temperatures.

A method suggested by Monsanto Chemical Co., is as follows:

5 mg Sopanox is added to, and uniformly mixed with, 5 g of soap. The treated soap is then placed in a 250-ml Erlenmeyer flask with a ground glass stopper. A 7-cm filter paper is cut into small pieces, saturated with a ml of distilled water, and added to the mixture in the flask. The flask is well shaken to mix the soap and paper and then

placed in an oven, maintained at 50°C. Each day the stopper is removed from the flask and the odor of the sample under test is noted. A blank test of the soap without Sopanox is run in parallel with the test indicated.

For determining the effect of discoloration by sunlight, the test is conducted by mixing thoroughly 20 g of freshly prepared soap with 0.1% by weight of Sopanox; 5 g of water is stirred into the soap and the mixture is pressed into a hard cake in a 2-inch crystallizing dish. It is then exposed to direct sunlight in a glass case and the discoloration observed, in comparison with soap containing no antioxidant.

Examination of Fatty Acids

Dissolve 4 oz soap in water in 800-ml beaker; add slowly 50 ml 50%-sulfuric acid. Heat on the steam bath until the clear fatty acids float on the surface, keeping the beaker covered with a watch glass. Syphon off the acid water, wash the fatty acids twice with hot water and decant them into a 100-ml beaker on top of the hot oven, to settle traces of water.

Place a beaker with funnel and filter paper in the oven, and when hot, filter the fatty acids. Determinations, such as titer, saponification and iodine numbers, qualitative and quantitative tests for rosin, etc., may now be made on the fatty acids.

Qualitative Test for Rosin

Liebermann-Storch Reaction

Dissolve several drops of fat in acetic anhydride, in a porcelain evaporating dish or on a dropping plate. Add a few drops of sulfuric acid of 1.53 specific gravity (17 ml of concentrated sulfuric acid is added to 19 ml of water). In the presence of rosin acids, a fleeting reddish-violet color appears. The test is very sensitive. Cholesterol, however, and some other substances give a similar reaction.

Quantitative Tests for Rosin

Wolff Method

The American Oil Chemists' Society's Official Method, Da 12–42, known as the *McNicoll* method, is somewhat more accurate than the Wolff method, but it requires a special reagent. For commercial work, the Wolff method is sufficiently accurate.

If the presence of rosin has been determined by the Liebermann-

Storch reaction, dissolve 3 g of fatty acids, in 20 ml of absolute ethanol. Add 10 ml of a solution of 1 volume concentrated sulfuric acid in 4 volumes of absolute alcohol, and boil for about four minutes under reflux condenser. Add to the liquid about five times its volume of a 7–10% solution of sodium chloride, and extract with ether. Shake out the aqueous portion two or three times with ether. Unite the ether solutions and wash with sodium chloride solution until the washings are neutral. Titrate the rosin acids with standard alcoholic sodium hydroxide solution. (1 ml N/alkali = 0.346 g rosin.) Calculate the percentage of rosin.

McNicoll Method *

Apparatus:

A glass flask connected, preferably by a ground glass joint, to a reflux condenser; a 150-ml flask of either the round-bottom or Erlenmeyer type, for carrying out the esterification; any suitable water-cooled reflux condenser.

Reagent:

An approximately N/5 solution of potassium hydroxide in neutral ethanol; naphthalene-β-sulfonic acid solution, made by dissolving 40 g Eastman grade or equivalent reagent in a liter of C.P. absolute methanol; phenolphthalein indicator; 0.5% solution in neutral redistilled alcohol.

Procedure:

Dissolve 5 g sample in 100 ml of hot water in a 250-ml Erlenmeyer flask. When solution is complete, add dilute sulfuric acid (1:1) in slight excess, using methyl orange as indicator. Insert a small funnel in the neck of the flask and heat to cause the fatty and rosin acids to separate as a clear layer. Cool to room temperature (20–25°C) and transfer to a separatory funnel, using small portions of ethyl ether to wash the flask. Draw off the acid layer into a second separatory funnel and shake the acid aqueous liquid with two 20-ml portions of ethyl ether. Dissolve the fatty acids in the ether used for washing the aqueous liquid, and shake with 10-ml portions of water until it is no longer acid to methyl orange. Unite the water portions used for washing and shake with 20 ml of ether. Wash the ether until the water is neutral to methyl orange.

* Federal Stock Catalog Method 70.3 *Rosin, McNicoll Method*

Unite the ether solutions, filter through anhydrous sodium sulfate on a filter paper and wash thoroughly with ether into the tared esterification flask. Evaporate the ether and finally heat the flask at 105°C for 15 minutes. When cool, weigh and calculate the percentage of fatty and rosin acids.

Pour 25-ml of naphthalene-β-sulfonic acid solution into the esterification flask. Add a few glass beads to ensure smooth boiling, attach the reflux condenser and boil for 30 minutes; also run a blank test using 25 ml of reagent. At the end of the boiling period, cool the contents of the flask, add 0.5 ml of phenophthalein indicator, and titrate immediately with standard alcoholic potassium hydroxide.

Calculations:

$$\%\ \text{Uncorrected rosin, original sample} = \frac{(S - B) \times N \times 34.6}{\text{Wt. of sample}} \qquad (R)$$

$$\%\ \text{Corrected rosin, original sample} = R - 0.01\ F \qquad (R_1)$$

$$\text{Rosin soda soap}\ \% = R_1 \times 1.064$$

In these calculations:

S, Titration of esterified fatty acids
B, Titration of blank
N, Normality of standard KOH solution
F, Percentage of total fatty acids in soap

If the true fatty acid soap is desired, subtract the rosin soap from the total anhydrous soap.

Where the rosin content is found to be less than 5%, the actual presence or absence of rosin should be checked qualitatively, by the Liebermann-Storch test.

Cooperative studies have shown that the McNicoll method gives results approximately 1% higher than the amount of rosin present. Consequently 1% should be deducted from the percentage of rosin found in the fatty acids, or 0.01F from the percentage found in the original soap.

Analysis of Potash Soap

With the exception of the determination of potash itself, the methods used here are similar to those employed in the analysis of soda soap.

Although a gel soap may appear to be made of potash, it is not always true. It is therefore essential to make a qualitative test for potash before going further.

Qualitative Test for Potash

Incinerate 1 g of soap, leach the ash with 10 ml of hot water, neutralize with acetic or hydrochloric acid, and add a few ml of 10% solution of sodium cobaltinitrite. A yellow precipitate on slight warming, indicates potash.

Quantitative Determination of Potash

The least expensive and easiest method for determining potash is with perchlorate, which the author has used for many years and found satisfactory for all practical purposes.

Incinerate 3 g of soap in a platinum crucible. Dissolve the ash in hot water, filter if necessary, and titrate with N/HCl, using methyl orange as indicator. Evaporate to dryness in a tall 300-ml beaker, on a hot plate. Dissolve the residue in hot water and add 5 ml of perchloric acid (20% solution). Evaporate again, to dense fumes. To ensure that all salts have been converted to perchlorates, dissolve as before, add 5 ml of perchloric acid and evaporate once more to dense fumes. *

To the perchlorates, add 25 ml of absolute ethanol, saturated with potassium perchlorate and 0.2% perchloric acid. Break up the crystals or dried solids with a stirring rod, allow to stand for one-half hour to ensure solution of all perchlorates, except potassium perchlorate. (This determination depends on the insolubility of $KClO_4$ in absolute alcohol.) Filter through a prepared Gooch crucible, washing the precipitate four times by decantation, with 25 ml of the alcoholic mixture. Transfer the precipitate to the crucible and wash it twice with 10-ml portions of ethanol solution and ether, 1:1. Dry for one hour in the oven at 120°–130°C. Cool in a desiccator and weigh. Dissolve the potassium perchlorate on the crucible with hot water, dry and weigh the crucible as before. The loss in weight represents the amount of potassium perchlorate.

$$\frac{KClO_4 \times 100}{3} = \%\ \text{Potassium perchlorate}$$

If a test of the soap shows the presence of sulfates, these must be precipitated with barium chloride and filtered off, before determining the potash.

* This test may be conducted on the filtrate from the total fats, but if considerable glycerol is present, it is best to start with the incinerated sample. This avoids the possibility of explosion when the glycerol is boiled down with perchloric acid.

Conversion Factors:

Multiply	by	to find
Potassium perchlorate	$\times$ 0.3399 =	K_2O
Potassium hydroxide	$\times$ 0.8394 =	K_2O
Potassium carbonate	$\times$ 0.6816 =	K_2O
Potassium chloride	$\times$ 0.6313 =	K_2O

Calculation of Soap Analysis

Total fat minus unsaponifiable $\times$ 0.97 = Fatty Anhydrides
Fatty anhydrides plus combined Na_2O = Anhydrous soap

In heavy silicated soaps, the carbonate should be determined by evolution and absorption in potash bulbs or similar device. When carbonate, silicate, and free caustic are subtracted from the total Na_2O, even in the absence of other alkalies, there may be more Na_2O left, than is necessary for the saponification of the fatty acids. This may be due to caustic soda, which is sometimes added to the soap in the crutcher to prevent the silicate from dropping out while the soap is cooling.

The true combined alkali may be determined by distilling off the alcoholic filtrate from the determination of free caustic, then decomposing the soap with normal sulfuric acid, and titrating the aqueous portion after removal of the fatty acids.

If the K_2O found gravimetrically, minus K_2O represented by chlorides, is equal to K_2O found by titration, the soap is made entirely with potash. If it is less, subtract from the total K_2O found by titration, the K_2O determined gravimetrically minus the K_2O as chlorides. The difference is recalculated to Na_2O. There are some soaps which are made partly with potash and partly with soda; e.g., shaving soap.

Analysis of Soap Powders

The analysis of soap powders follows the scheme outlined for soap. Soap powders, however, may contain several additional chemicals or builders not ordinarily found in soap. The better-class powders contain, besides sodium carbonate and silicate, also trisodium phosphate and other phosphates. The latter are found especially in the household detergents. It is therefore essential to make qualitative and quantitative tests for the various compounds.

Tetrasodium Pyrophosphate

Qualitative Test:

The reagent used is a solution of 5 g of zinc acetate and 20 ml of concentrated acetic acid in 100 ml of water.

Treat a 5-g sample of soap powder with 50 ml of hot 95% ethanol and filter. Wash the residue with hot ethanol and dissolve in hot water. Add concentrated acetic acid to the aqueous solution of the residue, until the *p*H drops to 3.8.

The *p*H may be determined in several ways: Using bromphenol blue —La Motte color standard 3 to 4; or if this is not available, Universal *p*H paper serves as a sufficiently close guide. The best, of course, is an electric potentiometer. Add 25 ml of zinc acetate slowly, with stirring. A white precipitate of zinc pyrophosphate indicates the presence of sodium pyrophosphate.

Sodium carbonate, bicarbonate, orthophosphate, borate, and silicate do not interfere.

Quantitative Determination (Method 1502.1)

Fed. Test Method Std. No. 536

1. *Reagents*
   - 1.1 Neutral ethyl alcohol (95%). Freshly boiled ethyl alcohol 95% or higher and neutral to phenolphthalein.
   - 1.2 Hydrochloric acid (approximately 0.5 N).
   - 1.3 Sulfuric acid (approximately 0.1 N).
   - 1.4 Zinc sulfate solution (7%). Dissolve 125 g of reagent-grade zinc sulfate ($ZnSO_4 \cdot 7H_2O$) in water and dilute to 1 liter, filter and adjust the *p*H to exactly 3.8 with a glass electrode titration assembly using 0.1 N sulfuric acid.
   - 1.5 Sodium hydroxide (approximately 0.5 N). Standardize the sodium hydroxide solution against reagent tetrasodium pyrophosphate ($Na_4P_2O_7$) prepared by recrystallizing the technical product three times from water, and igniting at 400°C to constant weight.
2. *Procedure:*
   - 2.1 Accurately weigh a sample containing an equivalent of approximately 1 g of tetrasodium pyrophosphate, and digest it on a steam bath with 200-ml of hot, freshly boiled, neutral alcohol. Filter through a filter paper or Gooch crucible with suction, and wash with hot alcohol until free from soap. Change receivers, placing a 400-ml beaker under the funnel, and wash the alcohol-insoluble portion remaining on the filter with hot water, until the washings are neutral to phenolphthalein. Cool and add sufficient distilled water, so that the resulting solution will just cover the electrodes of a glass-electrode titration assembly. Adjust the *p*H of the solution to exactly 3.8 with 0.1N sulfuric acid. Add 50 ml of zinc sulfate solution (7%), and allow 5 minutes for the reaction to become complete, as shown by the *p*H becoming constant. Titrate the liberated acid with

0.5N sodium hydroxide until a *p*H of 3.8 is again reached. This titration is a measure of the pyrophosphate content.

3. *Calculation:*

3.1 Calculate the percentage of tetrasodium pyrophosphate in the soap powder as follows:

$$\frac{\text{ml sodium hydroxide} \times F \times 100}{\text{Wt. of sample}} = \%\ \text{Tetrasodium pyrophosphate } (Na_4P_2O)_7$$

where F equals g of tetrasodium pyrophosphate equivalent to 1 ml of the sodium hydroxide solution used for the titration, as calculated by standardization against reagent tetrasodium pyrophosphate.

An alternate quantitative method for tetrasodium pyrophosphate which the author used for years in a soap laboratory with complete satisfaction is the following:

Treat 5 g of soap with 100 ml of hot 95% ethanol and filter. Wash the residue with hot ethanol to remove all traces of soap, and then dissolve it in hot water. Transfer an aliquot, containing about 0.1 g of tetrasodium pyrophosphate, to a 300-ml tall beaker and add concentrated acetic acid until the *p*H is 3.5. Adjust the volume to 75–100 ml, and add excess of 0.2 m zinc acetate solution, drop by drop, with constant stirring. Filter the precipitated zinc pyrophosphate and wash with zinc acetate reagent, diluted ten times with water. Ignite the filter and precipitate in a porcelain crucible, cool in a desiccator, and weigh.

As a choice of treatment, the precipitate may be dissolved in nitric acid, and after conversion to the orthophosphate by boiling the content of P_2O_5 may be determined in the usual way.

$$Zn_2P_2O_7 \times 0.8734 = Na_4P_2O_7$$

(The qualitative and quantitative methods were originally published in the *J. Am. Chem. Soc.*, 57:2284 (1935).

Trisodium Phosphate

Qualitative Test:

Prepare reagent as follows:

Dissolve 100 g of molybdic acid in 144 ml of concentrated ammonium hydroxide and 271 ml of water. Adding slowly with constant stirring, pour the solution so obtained into 489 ml of nitric acid and

1148 ml of water. Keep the mixture warm for several days and decant the clear solution.

Treat the sample with hot alcohol, as before, to free it from all traces of soap (and usually of any synthetic organic surfactant, if present). Filter and dissolve the residue in water. To a few ml in a test tube, add 3 ml of ammonium hydroxide and acidify with nitric acid. Heat to 65°C and add 10 ml of ammonium molybdate solution. Shake or stir the contents; a yellow precipitate indicates trisodium phosphate.

Quantitative Determination:

Treat 5 g of the product to be tested as in the previous determination, to free it of all traces of soap. Dissolve the alcohol-insoluble matter with water and filter into a 250-ml volumetric flask. Transfer a 10-ml aliquot to a 300-ml beaker and adjust the volume to 75 ml. Add 5 ml of ammonium hydroxide and 7 ml of concentrated nitric acid. The solution should be acidic at this point. Heat to 65°C and add 25 ml of ammonium molybdate slowly, with stirring, until the precipitate forms.

After an hour, when the ammonium phosphomolybdate in the sample under test has settled, it is filtered and the precipitate is washed with ice-water until two fillings of the filter paper require only 1 drop of normal alkali to produce pink color with phenolphthalein.

The filter and precipitate are transferred to a beaker, the yellow precipitate is slowly dissolved by flowing N/sodium hydroxide from a burette. The filter paper may be broken up so as to allow the precipitate to dissolve. Water is added and when no more yellow particles are visible, the excess sodium hydroxide is titrated back with N/sulfuric acid, using phenolphthalein as indicator.

$$\frac{\text{ml N/aOH} - \text{ml N}/H_2SO_4 \times 0.007124 \times 100}{\text{Wt. of sample}} = \%\ Na_3PO_4$$

Total Phosphates (Method 1501.1)

1.1 If a qualitative test has shown the presence of phosphates and their determination is desired, the matter insoluble in alcohol or the ash from the incineration of an original sample may be used.

2. *Reagents:*

2.1 Molybdate solution is the same as that used for determining trisodium phosphate.

2.2 Ammonium nitrate solution. Dissolve 100 g of phosphate-free ammonium nitrate in distilled water, and dilute to 1 liter.

2.3 Magnesia mixture. Dissolve 55 g of crystallized magnesium chloride ($MgCl_2 \cdot 6H_2O$) in water, add 140 g of ammonium chloride, make slightly ammoniacal to phenolphthalein, allow to stand overnight and filter. Acidify the filtrate with hydrochloric acid, add 5 ml of the acid in excess, and dilute to 1 liter.

2.4 Dilute ammonium hydroxide for washing. Dilute 100 ml of ammonium hydroxide (sp. gr. 0.90) to 1 liter.

3. *Procedure:*

3.1 Weigh out a 2 g sample of the alcohol-insoluble or ash, and proceed to remove the silica, if any, just as in the quantitative determination of sodium silicate, using nitric acid instead of hydrochloric. Boil the covered solution 30 minutes to convert other phosphates that may be present to orthophosphate, wash off and remove cover glass, and evaporate as described. Save the filtrate, Make up to 250 ml, concentrating if necessary. Pipette an aliquot corresponding to 0.5 or 1 gram of the original material, into a 250 ml beaker; add ammonium hydroxide in slight excess; and barely dissolve the precipitate formed with a few drops of nitric acid, stirring vigorously. Add about 15 g of dry ammonium nitrate or a solution containing that amount. Heat to about 60°C and add 70 ml of the molybdate solution for every decigram of phosphoric anhydride (P_2O_5) present.

3.2 Digest at about 65°C for an hour, and determine if the phosphates have been completely precipitated, by adding more molybdate solution to the clear supernatant liquid. Filter and wash with ammonium nitrate solution (see note 1). Dissolve the precipitate on the filter with ammonium hydroxide (1:1), and wash with hot water into a beaker, keeping the volume under 100 ml. Neutralize with hydrochloric acid, using litmus paper or bromthymol blue as an indicator, cool in ice, and from a pipette add slowly (about 1 drop per second), stirring vigorously, 15 ml of magnesia mixture for each decigram of P_2O_5 present. After 15 minutes, add 15 ml of ammonium hydroxide (sp. gr. 0.90) slowly whiles tirring. Let stand till the supernatant liquid is clear (2 hours is usually enough), filter, wash the precipitate with dislute ammonium hydroxide until the washings are practically free from chlorides, dry, burn first at a low heat (platinum should not be used), and then ignite to constant weight, preferably in an electric furnace, at 1050° to 1100°C; cool in a desiccator, and weigh as magnesium pyrophosphate ($Mg_2P_2O_7$).

Calculate and report the result as percentage of phosphoric anhydride (P_2O_5).

$$\frac{\text{Weight } Mg_2P_2O_7 \times 63.77}{\text{Weight of original sample}} = \% \text{ Phosphoric anhydride } (P_2O_5)$$

When the highest accuracy is not required, the molybdate precipitate and paper, after being thoroughly washed free from acid, may be transferred to a beaker, dissolved in a measured volume of standard sodium hydroxide solution and the excess titrated with standard acid solution using phenolphthalein as the indicator.

$$\frac{(\text{ml of NaOH} \times N)}{\text{Wt. of original sample}} - (\text{ml of HCl} \times N) \times 0.31 = \% \text{ phosphoric anhydride}$$

SYNTHETIC DETERGENTS

Presence of Synthetic Detergent

Qualitative Test: Method 5103

1.1 The presence of synthetic detergent shall be determined by dissolving approximately 5 g of the sample in 100 ml of warm distilled water. Add 25 ml of magnesium nitrate solution (20%). Stir and filter, collecting 50 ml of the filtrate in a 250 ml glass-stoppered graduated cylinder. Stopper and shake the cylinder vigorously for 10 seconds. A foam greater in volume than 10 ml and which persists for more than 1 minute, indicates the presence of synthetic detergent. Note. The absence of foam does not necessarily indicate the absence of synthetic detergent.

Another qualitative method is that of L. F. Hoyt:[11]

Small amounts of anionic, cationic, and nonionic surface-active agents can be detected by their solubilizing action on dilute aqueous suspensions of National Brilliant Oil Blue BMA.

If soap is present, the sample should be acidulated and the fatty acids filtered off, before testing the neutralized filtrate.

A 10-g sample is ground and digested with 50 ml of hot 50%-ethanol and filtered. The extract is evaporated to dryness, digested with 3 ml of distilled water and the solution is filtered.

As a reagent, 2 g of National Brilliant Oil Blue BMA is ground with 98 g of salt.

To the filtered aqueous solution of the extracted sample, 0.02 g of reagent is added. The solution is shaken and set aside for one hour. The presence of surface-active agents is shown by the formation of a blue solution.

To Differentiate Types of Surfactants

It is often necessary to know whether a surfactant is anionic, cationic, or nonionic. In one portion of an acidified test solution, the reaction of a surfactant with methyl yellow, shows a change from red to yellow or orange color only with nonionic and cationic surfactants.

In a second portion, the reaction of benzidine and metavanadate gives a dark green color only with nonionics; with cationics and sulfonated anionics, a purple. Then methyl yellow is added, which yields an orange color if a cationic surfactant is present. A list of 21 commercial surfactants and their manufacturers is included and their representative test results were given.[5]

Calculation of Anhydrous Salt-Free Material

In order to accomplish this (Method 6501), Method 5701 (Alcohol-soluble matter), Method 5601 (anhydrous salt-free, soda soap), and Method 6201 for Chlorides must be employed. These methods follow:

Alcohol-soluble and Alcohol Insoluble Matter (5701)

(Alcohol-soluble matter includes all the anhydrous soap, synthetic detergents, and some inorganic salts. Most of the sodium chloride will be included.) Applicable to soap and soap products including those containing synthetic detergents.

1. *Reagents:*

1.1 Neutral ethyl alcohol (95%) freshly boiled ethyl alcohol, 95% or higher and neutral to phenolphthalein.

2. *Procedure:*

2.1 Weigh 2 grams of the sample into a 250 ml beaker. Add 100 ml of neutral alcohol (95%), cover the beaker, and heat on the steam bath with frequent stirring until completely disintegrated. Let settle and filter the supernatant liquid through a counterpoised filter paper or a tared Gooch crucible with suction, into a 300-ml Erlenmeyer flask, retaining as much of the residue as possible, in the beaker. Repeat this extraction three times with 25-ml portions of hot neutral alcohol, each time retaining as much of the residue as possible in the beaker. Finally, evaporate any remaining alcohol and dissolve the residue in the smallest possible quantity of hot distilled water (5 ml is sufficient). Re-precipitate the alcohol-insoluble matter by slowly adding, while stirring vigorously, 50 ml of neutral (absolute) ethanol.

Note 1. Solution and reprecipitation of alcohol-insoluble matter is necessary for complete separation from alcohol-soluble matter.

2.2 Heat the solution to boiling on the steam bath, filter and transfer the precipitate quantitatively to the Gooch crucible or filter paper, washing several times with neutral ethyl alcohol (95%). Combine the alcohol filtrate and washings in a tared 250 ml beaker, evaporate on the steam bath, and then dry to constant weight at 105°C ± 2°. Calculate total alcohol-soluble matter.

2.3 Reserve the beaker and contents for the determination of free fatty matter (Method 6101). Reserve the paper or the Gooch crucible and contents, without drying, for the determination of matter insoluble in water (Method 5801) and total alkalinity of matter insoluble in alcohol (alkaline salts) (Method 5901).

Anhydrous, Salt-Free, Soda Soap (5601)

(This method determines the total anhydrous, salt-free, soda soap in the sample. Applicable to soaps and soap products containing synthetic detergents.)

1. *Apparatus:*
   1.1 Separatory funnel, 250 ml capacity.
   1.2 Separatory funnel, 500 ml capacity.

2. *Reagents:*
   2.1 Neutral ethyl alcohol (95%) as previously described.
   2.2 Methyl orange indicator: Prepare a 0.1% solution in water.
   2.3 Hydrochloric Acid (approximately 0.5 N).
   2.4 Hydrochloric Acid (1:1): Slowly add 1 part of HCl (sp. gr. 1.19) to 1 part of distilled water.
   2.5 Petroleum ether.
   2.6 Sodium sulfate, anhydrous, reagent-grade.
   2.7 Phenolphthalein indicator 1% as previously described.
   2.8 Standard sodium hydroxide (approximately 0.1N, accurately standardized.
   2.9 Sodium chloride solution: Saturated solution in distilled water.

3. *Procedure:*
   3.1 Weigh a sample (sufficient to yield 1 to 5 g of fatty acids, into a tared 250 ml beaker. Add 25 ml of water and 25 ml of neutral 95% ethanol, and warm on the steam bath until solution is complete. Cool, add 5 drops of methyl orange indicator, and titrate with approximately 0.5 N hydrochloric acid to a pink color. Add 5 ml of HCl (1:1) in excess.
   3.2 Transfer the contents of the beaker to a 250-ml separatory funnel. Wash the beaker alternately with equal parts of hot distilled water and hot ethanol 95%, adding the washings to the separatory funnel. Add sufficient sodium chloride solution

to make 160 ml in the funnel. Wash the beaker with a small amount of petroleum ether to remove any traces of fatty acids and fatty matter and add to the funnel. Add 50 ml of petroleum ether and allow to stand for one-half hour without shaking. Remove the greater part of the fatty acids by drawing off the petroleum ether layer as closely as possible into a 500-ml separatory funnel. Repeat the extraction five more times with petroleum ether, using 50-ml portions and shaking thoroughly each time.

Note 1. If an emulsion appears at this point, it may be broken by the addition of about 10 g of anhydrous sodium sulfate.

3.3 Combine the petroleum ether extracts and wash with small portions of distilled water until the water washings are no longer acid to methyl orange indicator. Dry the combined, washed petroleum ether extracts with anhydrous sodium sulfate, and filter through paper into the original tared 250 ml beaker. Wash the separatory funnel with two small portions of petroleum ether, filtering and adding the washings to the beaker.

3.4 Evaporate the petroleum ether extract on the steam bath until the odor of petroleum ether is no longer perceptible. Then place the beaker in an oven at 100° to 105°C for five minutes to ensure complete removal of the solvent. Cool and weigh as fatty acids plus fatty matter.

3.5 Dissolve the fatty acids and fatty matter in 50 ml of neutral alcohol (95%) with warming. Add phenolphthalein indicator and titrate with standard sodium hydroxide solution to a pink end point. Evaporate off the alcohol, dry to constant weight, and calculate to obtain soda soap plus fatty matter (G). Make a blank test on the sodium hydroxide solution for neutral salts and correct the result accordingly. (See Method 401, pars. 1.6 and 3.)

4. *Calculation:*

4.1 Calculate the percentage of anhydrous, salt-free, soda soap as follows:

$$A = G - F$$

G may be calculated as follows:

$$G = \frac{(VN \times 0.022) + E \times 100}{W}$$

Where A = % anhydrous, salt-free soda soap
G = % soda soap plus fatty matter (3.5) or calculated
F = % fatty matter. (See Method 6101, par. 3.4)
V = ml, sodium hydroxide solution used in titration (3.5)
N = normality, sodium hydroxide solution
E = grams of extract (3.4)
W = grams of sample (3.1)

Fatty Matter

(This is a resumé of Method 6101) The alcohol-soluble matter (obtained as specified in Method 5701) is dissolved in a mixture of 25 ml of ethanol and 25 ml of water, warming if necessary. The solution is transferred to a separatory funnel and shaken out with petroleum ether. The combined petroleum ether extracts are washed four times with 10-ml portions of N/5 sodium hydroxide. The petroleum ether extracts are evaporated on the steam bath, dried at 100–105°C for half hour, cooled in a desiccator and weighed. The percentage of fatty matter is calculated.

Chlorides

The alcoholic solution from the determination of fatty matter is saved for Method 6201.1, *Chlorides.*

To this solution is added 25 ml of 20% magnesium nitrate, warmed to precipitate and coagulate soaps. If the solution is alkaline it must be neutralized with N/sulfuric acid. It is then titrated with N/10 silver nitrate, using potassium chromate as indicator. A blank, using the same volume of water and magnesium nitrate as in the test, is titrated with silver nitrate.

Calculate the percentage of chloride (NaCl) in alcohol-soluble matter as follows:

$$C = \frac{(V_1 - V_2)N \times 5.85}{W}$$

Where C = % NaCl in alcohol-soluble matter
C_1 = % KCl (if any) in alcohol-soluble matter
V_1 = ml $AgNO_3$ required for titration of sample
V_2 = ml $AgNO_3$,, ,, ,, ,, blank
N = Normality of $AgNO_3$
W = Weight of sample

Having made all of these determinations, the synthetic detergent calculated according to Method 6501 is

1. *Determination by Difference:*
   1.1 Calculate the percentage of anhydrous, salt-free synthetic detergent as follows:

$$D = A - (W + F + C)$$

where D = % anhydrous, salt-free, synthetic detergent
A = % alcohol-soluble matter (Method 5701)
W = % anhydrous, salt-free soda soap (5601)
F = % fatty matter (Method 6101)
C = % chlorides as (NaCl) in alcohol-soluble matter (Method 6201)

TESTS FOR BIODEGRADABILITY

Though foam from the faucet and suds in the stream, pose definite problems, which are not attributable entirely to the presence of conventional surfactants, the change-over to linear alkylates is calculated to mitigate the effect insofar as detergents can be blamed.[8]

Researchers on detergents employ several analytical approaches to determine the degree of biodegradation for syndets. In industry's three most widely used tests, the straight-chain surfactants, show practically complete degradability in minimum time.

Quoted from Conoco's literature, the following is a brief and concise description of the three tests:

Activated Sludge Method

Laboratory-activated sludge simulates conditions in large-scale sewage disposal operations. Both continuous (dynamic) and batch (static) methods utilize sludge from waste-disposal plants as the bacterial culture. The batch procedure, sometimes termed *semi-continuous*, or *fill and draw*, has gained wide acceptance, because it requires only simple and inexpensive equipment even though it offers limited throughput and does not accurately duplicate conditions in a sewage plant.

In this test, the sample of detergent is added to water containing a synthetic nutrient medium and activated sludge, in a vertical cylindrical chamber. Air bubbling through the liquid, agitates the mixture. After a 23-hour testing period, the air is turned off to allow the sludge to settle. The supernatant liquid is then sampled and analyzed. This company further states that Conoco's biodegradable detergents average 95–100% degradation over a weekly series of these one-day tests. These results correlate with data on biodegradability from the unique, continuous activated-sludge system in Conoco's Ponca City, Oklahoma, research center. This dynamic laboratory unit closely simulates conditions found in full-scale sewage-disposal plants.

River Die-away Method

River water is added to a sample of detergent solution in a clean glass jar. Covered with a loose top, the liquid incubates at rest in the dark. Activating a magnetic stirrer for one minute gives adequate mixing before periodic sampling. Conoco normally runs this test 20 to 30 days to determine the degree of biodegradability of various surfactant materials. In general, soft detergents achieve essentially complete degradation in 5 to 20 days under these conditions.

Shake-Flask Method

After surfactant solution is measured into a shake-flask, the sample is sterilized in an autoclave, cooled, and inoculated with an adapted mixed microbial culture 72 hours old. Loosely capped after addition of environmental medium, the liquid incubates at 25°C during continuous agitation on a rotary shaker. Biodegradable detergents generally take two to five days to degrade in this test.-

TABLE 9:1

MOISTURE TESTS ON SOME POPULAR LIQUID DETERGENTS

Vel Dishwashing Compound	67.0% H_2O	
Chiffon ,, ,,	57.0	
Ivory Liquid	42.0	
Joy for Dishes	44.0	*p*H 6
Wisk	54.0	USP 2 859 182
P&G Pink Lotion Dishwashing Comp.		USP 2 618 608
Mr. Clean	65.0	*p*H 10
Lux Pink Lotion Dishwashing Comp.	53.28	
Top Job	69.67	*p*H 9

Analysis of residual surfactants lends itself to either manual or automated analytical procedures. The best of today's automated analytical systems provides high accuracy and prodigious throughput as well. The trace quantities of active detergent components present in treated samples (20 ppm down to 0.5 ppm) have required the development of sensitive and reproducible techniques. Conoco's chemists and microbiologists use standard methylene-blue colorimetry for determining anionic compounds. For nonionics, a special cobaltothiocyanate colorimetric procedure yields uniformly accurate results.[7] Upon request, Conoco will gladly furnish detailed information on its tests and analytical techniques for determining detergent biodegradability.

ANALYSIS OF GLYCEROL

Dichromate Method

Weigh 20 g of bar soap or 15 g of flakes or powdered soap. Dissolve in 100 ml of water and add sufficient sulfuric acid (1:1) to separate the fatty acids. Heat on a steam bath until the fatty acids are clear, keeping the beaker covered with a watch glass. Cool and filter into a 250-ml volumetric flask. Neutralize the filtrate with 10% sodium hydroxide solution and add basic lead acetate (dry powder), rotating the flask after each addition, until no further precipitate forms. Fill to mark, shake, filter and transfer 25-ml aliquot into a 300-ml Erlenmeyer flask. Add a few ml of dilute sulfuric acid to precipitate the lead salts.

Weigh out in a small evaporating dish exactly 2 g of C.P. potassium dichromate crystals and wash this into the Erlenmeyer flask. When the dichromate has dissolved, add 50 ml of sulfuric acid (1:1), cover with a small beaker, and keep in a steam bath for two hours. Cool and titrate with a standard solution of ferrous ammonium sulfate, running a blank with 2 g of dichromate and 50 ml of sulfuric acid 1:1, and placing this in the steam bath with the Erlenmeyer flask that contains the sample.

Where such determinations are made infrequently, powdered ferrous ammonium sulfate may be used. It is added from a weighing bottle, directly into the cooled Erlenmeyer flask, a little at a time, to slight excess, titrating back with a 1% solution of potassium dichromate. A little skill must be developed in recognizing the change in color of the dichromate, which will be a muddy green or even dark brown after boiling, depending on the amount of glycerol present. Upon adding excess ferrous ammonium sulfate, the solution turns bluish green. A freshly prepared 0.1% solution of potassium ferricyanide is used as indicator, on a dropping plate. Excess of iron salt is indicated by a deep blue color (Turnbull's blue) when dropped over a drop of potassium ferricyanide. The end point is indicated by a light green color.

Titrate a blank containing 2 g of potassium dichromate by adding ferrous ammonium sulfate from a weighing bottle, making sure that all the crystals are dissolved after each addition, until a drop on the plate with potassium ferricyanide turns blue. If the end point is approached cautiously, one may obtain a slight green color and then, the 1%-dichromate need not be used to titrate back any excess. We find that exactly 16.12 g of ferrous ammonium sulfate is required. Dividing 2.00 by 16.12 = 0.1241 g dichromate, which is equivalent to 1 g of ferrous ammonium sulfate.

On titrating the sample, we find that 11.597 g of ferrous ammonium sulfate and 2 ml of 1%-dichromate are required to bring back the excess to a light green end point. Consequently, we have used 2.02 g of dichromate:

$$11.597 \times 0.1241 = 1.439 \text{ g potassium dichromate}$$
$$2.02 - 1.439 = 0.581 \text{ g dichromate}$$

actually used to oxidize the glycerol in 2 g of soap.

As 1 g of dichromate oxidizes 0.13411 g of glycerol,

$$\frac{0.581 \times 0.13411 \times 100}{2} = 3.85\% \text{ glycerol}$$

This analysis was made on a half-boiled potash soap, which contained about 35% of vegetable oils. The result is very close to the theoretical.

Acetin Method for Crude Glycerol

It has been found that in the analysis of pure glycerol either by the dichromate method or by the acetin method, the results are identical. In the analysis of crude glycerol, however, the dichromate method gives higher results. Consequently, by agreement, between committees of American, British, French, and German chemists, it has been decided that for crude glycerol, which contains not in excess of 60% of water, the acetin method gives the truer result.

The highest grade of acetic anhydride should be employed. Chemically pure sodium acetate should be fused again carefully, without charring, in a platinum or silica dish, quickly ground up, and kept in a stoppered bottle in the desiccator. Other reagents used in this analysis are *normal sodium hydroxide*, free from carbonate; *normal sulfuric acid*, carefully standardized; 0.5 g of phenolphthalein dissolved in 50 ml of alcohol and neutralized.

Into a thoroughly cleaned and dry narrow-mouth round-bottom flask, weigh rapidly 1.25 to 1.50 g of glycerol. A Grethan or Lunge weighing bottle will be found useful. Add 3 g of fused sodium acetate and 7.5 ml of acetic anhydride. Connect the flask with an upright Liebig condenser, the inner tube of which should not be over 50 cm long. The flask may be connected to the condenser by either a ground glass joint or a rubber stopper, which has had a preliminary treatment with hot vapors of acetic anhydride. Heat for one hour, taking precautions to keep the salts from drying on the sides of the flask during boiling.

Allow the flask to cool and add 50 ml of distilled water at 80°C through the tube of the condenser. By giving the flask and condenser a rotary motion, the salts may be brought into solution. Caution must be observed so as not to loosen the stopper until the flask is quite cool. When cool, wash down the inside of the condenser tube with water, loosen the stopper and wash it down, then filter the contents of the flask, though an ashless filter paper into a liter flask. Wash thoroughly with cold distilled water free from carbon dioxide. Add 2 ml phenolphthalein solution and add normal sodium hydroxide from a burette, until the solution in the flask has been neutralized to a faint, pinkish-yellow throughout.

The neutralization must be done very carefully while the solution in the flask is being swirled. Now add 50 ml of normal sodium hydroxide. Boil gently for 15 minutes with an air condenser. Then cool the flask and titrate with normal acid until the pinkish-yellow color previously chosen for the end point is reached. From the volume of normal sodium hydroxide consumed, the percentage of glycerol, including acetylizable impurities, may be calculated, after making correction for the following blank test. One ml of normal sodium hydroxide is equivalent to 0.03069 g glycerol.

Blank Test

The acetic anhydride and sodium acetate may contain impurities that affect the result. A blank determination, is therefore, run exactly as for the glycerol, except that no glycerol is added. It is not necessary to filter the solution of the melt, but sufficient time must be allowed for the hydrolysis of the acetic anhydride before neutralization.

When the solution has been neutralized to a pinkish-yellow, 10 ml of normal sodium hydroxide is added and the solution is boiled 15 minutes, as in the test for glycerol. When cool, the excess is titrated with normal acid. The amount of sodium hydroxide absorbed may be calculated to glycerol equivalent and then deducted from the percentage found in the first test.

Glycerol Value of Acetylizable Impurities

The residue at 160°C is found by dissolving 10 g of sample in distilled water, in a 100 ml volumetric flask. The solution is made alkaline to 0.2%-sodium oxide, with sodium carbonate, or reduced with hydrochloric acid, if too strongly alkaline. It is made up to the 100 ml mark and thoroughly mixed.

From this solution, 10 ml is transferred to a Petri dish and evaporated to dryness on the steam bath. The Petri dish is then placed in the oven, at 140°C until most of the glycerol has volatilized. The residue is taken up with water, evaporated slowly, and then placed in the oven at 160°C, where it remains for an hour. The solution of the residue is evaporated and heating at 160° is repeated. Finally it is cooled in a desiccator and weighed. The process is repeated until the loss in the oven is 1 to 1.5 mg an hour.

The residue is taken up with a few ml of water and transferred to the acetylating flask. It is dried in an oven and acetylated and its glycerol equivalent determined.

Calculation of Actual Glycerol Content

Determine the apparent glycerol in the sample; that is, the first test minus the value of the blank test. Determine the total residue at 160°C, then determine the acetyl value of the residue in terms of glycerol, and deduct this from the apparent glycerol content.

British Standards

A series of British standards for glycerol has been published in one volume as B.S. 2621-5, now fully revised.[13] It covers soap lye crude glycerol, hydrolyzer, saponification crude, technical, dynamite, and chemically pure glycerol. There are many changes for sampling and analysis. In determining content of glycerol, the method specified—an amended periodate procedure—has displaced the acetin method, similar to the method adopted by the International Union of Pure and Applied Chemistry. Copies of B.S. 2621-5 may be obtained from BSI Sales Branch, 2 Park St., London W.1., or from the American Standards Association.

REFERENCES

1. Official and Tentative Methods of the American Oil Chemists' Society, 1964.
2. Polenske Number: Official Test Method C d 2 .38 American Oil Chemists' Society.
3. Federal Test Methods, Standard No. 536: *Soap and Soap Products (Including Synthetic Detergents) Sampling and Testing;* General Services Administration, Business Service Center, Washington 25, D.C.
4. Analysis of Synthetic Detergents: Symposium, (9):1914 *Ind. Eng. Chem.* 46.
5. Seymour Greenberg: Differentiation tests for anionic, nonionic and cationic surfactants; *Chemist-Analyst*, 51:11–14 (1962). (The *Chemist-Analyst* is published by J. T. Baker Chemical Company, Phillipsburg N.J.)
6. R. Bernstein and H. Haftel, Determination of lanolin in powdered hand soap, *Soap San. Chemicals*, 30:48, November, 1954.
7. Irving R. Schmolka and J. W. Hensley, Detergent testing of nonionics, *Soap Chem. Specialties*, August, 1965.
8. P. J. Weaver, Biodegradability field tests and test methods, *Pro-*

ceedings Mid-Year Meeting, Chemical Specialties Manufacturers Association, 51:121–30 (1965).
9. An Apparatus for comparison of foaming properties of soaps and detergents *Oil and Soap*, May 1941, p. 99; USP 2 315 983 to Ross and Miles.
10. Test for Detergent Efficiency, *Oil & Soap* 20:9 (1943).
11. L. F. Hoyt, *J. Am. Oil Chemists Soc.*, 24:54 (1947).
12. Standard Methods for the Examination of Water and Waste Water (11th Ed.) p. 246; *Amer. Pub. H. Assoc.*
13. *British Standards:* 2621-5, Periodate Method for Determination of Glycerol.

Index

S

www.ingramcontent.com/pod-product-compliance
Ingram Content Group UK Ltd.
Pitfield, Milton Keynes, MK11 3LW, UK
UKHW041640031026
15587UKWH00001B/397

9 780820 602325